Multiple Criteria Decision Making

This book series focuses on the publication of monographs and edited volumes of wide interest for researchers and practitioners interested in the theory of multicriteria analysis and its applications in management and engineering. The book series publishes novel works related to the foundations and the methodological aspects of multicriteria analysis, its applications in different areas in management and engineering, as well as its connections with other quantitative and analytic disciplines. In recent years, multicriteria analysis has been widely used for decision making purposes by institutions and enterprises. Research is also very active in the field, with numerous publications in a wide range of publication outlets and different domains such as operations management, environmental and energy planning, finance and economics, marketing, engineering, and healthcare.

This series has been accepted by Scopus.

Constantin Zopounidis • George Fakotakis

Multi-Criteria Methodology for the Dynamic Development of Tourist Areas

A Guide to Sustainable Tourism Strategies

Springer

Constantin Zopounidis
School of Production Engineering and Management
Technical University of Crete
Chania, Greece

George Fakotakis
CIHEAM Mediterranean Agronomic Institute of Chania
Crete, Greece

ISSN 2366-0023 ISSN 2366-0031 (electronic)
Multiple Criteria Decision Making
ISBN 978-3-031-95313-2 ISBN 978-3-031-95314-9 (eBook)
https://doi.org/10.1007/978-3-031-95314-9

This Springer imprint is published by the registered company Springer Nature Switzerland AG.
The registered company address is: Gewerbestrasse 11, 6330 Cham, Switzerland

Preface

This book explores the evolving dimensions of visitor satisfaction in protected areas, with a focus on Samaria National Park in Crete, Greece. The study spans five consecutive years and includes the period of the COVID-19 pandemic, providing a unique opportunity to examine how visitor expectations and perceptions shifted in response to extraordinary global circumstances.

The motivation behind this research was both academic and practical. Protected areas are critical not only for conservation but also for providing meaningful experiences to visitors. Understanding the factors that influence satisfaction—and how these factors change over time—is essential for effective management and sustainable tourism development. This work aims to contribute to this understanding by applying structured methods to capture and analyze visitor feedback, highlighting both stable criteria and emerging trends.

The book is intended for researchers, practitioners, and policy-makers in the fields of environmental management, sustainable tourism, and protected area governance. It may also serve as a resource for graduate students interested in visitor experience assessment and post-pandemic tourism behavior.

The analysis draws on original field data and models that emphasize both the subjective and contextual nature of satisfaction. While rooted in the specific case of Samaria National Park, the findings offer broader insights relevant to similar protected areas globally.

We hope this work will not only inform ongoing efforts to improve visitor experience in protected areas but also encourage further research into the dynamic relationship between people and nature in times of both normalcy and disruption.

Chania, Greece — Constantin Zopounidis
Crete, Greece — George Fakotakis

Acknowledgments

This book is the result of a collaborative effort, and we are deeply grateful to all those who supported us along the way.

We would like to thank the Forestry Service of Chania and the Management Body of Samaria National Park for granting us access to the site and for their consistent cooperation throughout the study period. We are especially appreciative of the staff and rangers of the Park, whose assistance and on-the-ground perspectives were invaluable during our fieldwork.

We are sincerely grateful to the visitors who participated in the surveys and generously shared their time and opinions. Their responses form the backbone of our analysis.

Our appreciation also goes to our colleagues at CIHEAM-MAICh and Technical University of Crete, whose feedback and support enriched both the academic and practical dimensions of this work.

We would also like to express our heartfelt thanks to our families and friends for their continued encouragement and patience during the many phases of research and writing.

Competing Interests The authors have no competing interests to declare that are relevant to the content of this manuscript.

About the Book

This book investigates the factors influencing visitor satisfaction in protected areas, using Samaria National Park in Crete, Greece, as a case study. Spanning five consecutive years—including the period of the COVID-19 pandemic—the study explores how visitor preferences, behaviors, and perceptions evolved over time in response to both environmental and societal changes.

Using structured survey data and multi-criteria analysis methods, the research identifies key dimensions of satisfaction such as safety, available time, landscape, staff performance, and the quality of infrastructure. The results highlight both stable and shifting priorities among visitors, with notable changes in attitudes toward health-related safety, crowding, and travel motivation during the pandemic years.

The study provides actionable insights for park managers and policy-makers, emphasizing the need for adaptive management strategies that account for emerging visitor expectations and sustainability goals. While grounded in the context of Samaria National Park, the methodology and findings are broadly applicable to other high-traffic natural areas seeking to balance conservation with visitor experience.

Contents

Chapter 1
Introduction

1.1 The Samaria National Park

The tourism sector is a very important area of economic activity in Greece. Among the most well-known tourist destinations in Greece during the summer season is Crete, attracting hundreds of thousands of visitors, both Greeks and foreigners. In the Chania region, there are natural areas of high value and exceptional beauty protected by law, such as the Samaria National Park and the Balos Lagoon. These areas have gained international recognition over time and are a major reason why many visitors choose Chania for their holidays.

This study focuses on the Samaria National Park, which was created around the Samaria Gorge and is the most recognized and highly awarded protected area in Chania. The park's core includes 48.480 square kilometers of forested areas, meadows, and former agricultural lands, now designated as state-owned property. The surrounding area consists of forests, woodlands, meadows, and agricultural lands. However, there is an ongoing dispute regarding their ownership between the state, municipalities, and local residents.

The gorge itself lies within the White Mountains range. Its northern entrance, called Xyloskalo, is located at the Omalos Plateau at an altitude of 1227 meters. The gorge ends at the Libyan Sea after a 16-kilometer route leading to the village of Agia Roumeli. The length of the route within the boundaries of the National Park is 12,800 meters from the northern to the southern entrance, with an additional 3200 meters to the sea.

For many years, the management of the Samaria National Park was a shared responsibility between the Chania Forest Directorate and a special Management Body established in 2003. The board of the Management Body met once in 2004 at the facilities of MAICh, but its activities ceased thereafter. Ultimately, the body resumed operations in April 2008, primarily funded by the European Union. Among other functions, its responsibilities included managing the park, assisting local

C. Zopounidis, G. Fakotakis, *Multi-Criteria Methodology for the Dynamic Development of Tourist Areas*, Multiple Criteria Decision Making,
https://doi.org/10.1007/978-3-031-95314-9_1

authorities in enforcing regulations for its protection, collecting environmental data and processing them into databases, conducting research within the park, and implementing technical projects. It was also responsible for carrying out European Research Projects, such as the "Management and Operation of the Samaria National Park" for the 2008–2009 period with a total budget of €456,082.68, and "Protection and Management of the Biodiversity of the Samaria National Park" for the 2010–2015 period with a total budget of €4,460,615.79.

The ultimate goal of the Management Body was to evolve into an autonomous entity capable of assuming full responsibility for managing the park. This became possible only in 2022, when the Natural Environment and Climate Change Organization (O.FY.PE.K.A.) absorbed the Management Body and took over the management of the Samaria National Park, while protection responsibilities remained with the Chania Forest Directorate.

The uniqueness of the area's natural environment has been recognized worldwide, which is why the Greek State officially established the National Park in 1962. Until then, the area was inhabited by permanent residents, particularly in the old village of Samaria, located in the middle of the gorge. For the creation of the park, residents' properties were expropriated, and they were relocated. A state of strict protection was imposed, with stringent operational regulations. To preserve the unique natural environment, which includes endemic species of flora and fauna such as the Cretan wild goat, all human intervention was prohibited. At the same time, guard stations were established at key points of the park, especially at its entrances, to monitor and enforce protection measures. After its establishment, the park gradually attracted the attention of visitors and became a popular and renowned tourist destination.

Over the years, it has received various awards and international distinctions. These include being designated a UNESCO Biosphere Reserve in 1981, receiving the European Council's Diploma for Protected Areas in 1979, and being named a Biogenetic Reserve. It has also received the national diploma for natural protection, is protected under the Barcelona Convention, is listed as an Important Bird and Biodiversity Area of Greece, and is recognized as a site of exceptional natural beauty and a wildlife refuge. It is also part of the European Union's Natura 2000 network and generally holds the most significant international distinctions in Greece (Papavasileiou, 2008).

Human activity in the area dates back to ancient times. The Samaria Gorge is the only gorge in Chania that has been continuously inhabited, partly because it is accessible, navigable, and has abundant water sources. The first signs of human presence date back to the sixth century BC. Over the years following the park's establishment, various artifacts, coins, weapons, idols, and religious worship sites have been discovered. At the park's southern exit in ancient times stood a city named Tarra. Due to the inhospitable natural environment, the city was practically isolated from the rest of Crete and relied on its own resources for survival. It had strong fortifications to protect itself from pirates and drew fresh water from the Samaria Gorge. Communication with the outside world was done from the sea, and it served as a stopover for trade ships sailing to and from North Africa. Tarra was part of a federation of Cretan cities called the "Commonwealth of the Mountains" and even minted its own coin, depicting a bee on one side and a wild goat on the other. However, due to the catastrophic consequences of the major earthquake in 365 AD, the city was gradually abandoned (Papavasileiou, 2008).

The Samaria Gorge played a prominent role in every major historical event in Crete. During the late Ottoman period, when Ottoman rule began to collapse, the area, along with the entire island, fell into a state of anarchy due to the long-standing hostility between the Christian and Muslim populations. During the Daskalogiannis Revolution, the first major uprising against the Ottomans in 1770, and in every subsequent revolution, the gorge served as a refuge for civilians and revolutionaries. To control the area and suppress the 1866–1867 revolution, the Ottomans designed and built a network of forts known as "koule", each strategically located to communicate with others using smoke signals when help was needed. These forts remain visible along the route today, although most pf them are in poor condition (Papavasileiou, 2008).

In modern times, the most notable human activity in the wider Samaria Gorge area occurred in the villages of Samaria and Agia Roumeli. The village of Samaria is located 7 kilometers from the northern entrance of the gorge at an altitude of 350 meters. In the past, there were two settlements: an upper and a lower village, situated a short distance apart. Permanent settlement in the area likely began after the period of ancient Tarra. Living conditions were difficult, with limited arable land, high humidity, restricted sunlight, and isolation from the rest of the world. Nevertheless, the inhabitants managed to survive, settle permanently, and achieve self-sufficiency. They built stone houses and walls to define their properties, cultivated all available land, raised animals, and hunted game such as the Cretan wild goat. Water was sourced from the river running through the Samaria Gorge or various springs, with accounts of a wooden channel being constructed to transport water directly from a spring to the village. As mentioned earlier, the village of Samaria was expropriated in 1962 as part of the creation of the National Park. At that time, the upper settlement had 12 houses, while the lower settlement had 7 houses. While the upper settlement still had permanent residents, the lower one had been abandoned earlier, in the 1920s, leaving its houses in worse condition. The last residents relocated in 1965. Some houses in the village were repaired and preserved to be used by park administrators for staff accommodations and for educational and research activities (Papavasileiou, 2008).

The village of Agia Roumeli is situated on the coast at the end of the gorge. It played a significant role in Cretan history and was destroyed by the Ottomans multiple times. In the past, locals lived in the old, now-abandoned village of Mesa Agia Roumeli, located approximately one kilometer north of the new village. A flood in 1954 destroyed much of the old settlement. As a result, the Ministry of Social Welfare expropriated an area of 21,500 square meters to aid the flood victims, leading to the establishment of the new village. When the properties of the Samaria village within the gorge were expropriated for the creation of the National Park, some of the displaced residents also chose to move to Agia Roumeli. This, among other factors, demonstrates the close connection between life in Agia Roumeli and the Samaria Gorge. Since there is no road connecting Agia Roumeli to the rest of Crete, traversing the gorge is one of the few ways to reach the village. In the past, the local economy was mainly focused on the primary sector, with an emphasis on self-sufficiency. All families in the area kept flocks of sheep and goats, raised for meat, wool, and milk. The milk was processed into cheese and yogurt in local dairies, while wool was used to produce textiles. These products met the needs of the residents and were sometimes sold in other villages or the city of Chania. Agriculture was also an important part of the local economy, primarily involving the cultivation of olive trees and the production of olive oil. Evidence of this includes the significant number of trees still present in the park. The processing was done in local mills, one of which remains in good condition in Agia Roumeli. Other crops, such as wheat, vegetables, vines, and citrus fruits, were grown on a much smaller scale. Beekeeping was also very popular, with many families producing their own honey using manual equipment. The exploitation of forest resources was another key activity for the area's inhabitants. Logs were cut from the gorge, transported via the river to Agia Roumeli, and then shipped elsewhere for sale. Tree bark was also collected and sold for pitch production. Lime and charcoal were produced in local kilns and transported to Agia Roumeli for sale. For processing primary-sector products, water from the river was used to power mills for grinding grain, olive presses, and timber cutters. Fishing in the Libyan Sea and hunting in the area were also widespread. The Cretan wild goat was particularly popular as game, drawing hunters from other regions. The impact of hunting on the wild goat population was so significant that conservation measures were implemented as early as the 1930s. After the creation of the National Park, the expropriation of the area, and the relocation of the last residents, most human activity within the park's core ceased. Livestock farming, agriculture, hunting, and almost all other human activities were prohibited to preserve the untouched natural environment.

The establishment of the park, the international recognition it gained through various awards, and its transformation into a well-known tourist destination also affected the wider area. Today, most of Agia Roumeli's economic activity revolves around tourism, highlighting its close connection to the gorge. Once-dominant primary-sector activities have significantly declined, with very few people now earning a full-time living from them. The self-sufficient lifestyle has disappeared, and the village is characterized by what many describe as a "dual life." Very few residents live there permanently, with most only coming during the summer season. Restaurants and hotels have been built to accommodate visitors. The number of available beds increased from 106 in 1981 to 500 in 2008, although official figures from the Greek National Tourism Organization reported only 210 beds in 2008. Tourism development in Agia Roumeli is, however, marked by its inability to retain visitors. Most tourists arrive after hiking through the Samaria Gorge and either leave the same day or stay for only a short time. Although the area has the potential to become an independent tourist destination rather than merely the exit point of the Samaria Gorge, development efforts and marketing attempts have yet to yield significant results in this direction (Papavasileiou, 2008).

The development of the Samaria Gorge area, following the establishment of the National Park, was better planned on the other hand. The local environment has been preserved with minimal interventions along the trail. The trail itself is maintained annually, with wooden railings installed where necessary to blend with the natural surroundings. In areas prone to rockfalls, nets have been placed to prevent accidents, while signs along the trail inform visitors about points of interest. Along the route, park staff are stationed in key stations to handle emergencies and protect the area. Various buildings along the path have been restored and are being used by park staff, to house medical facilities, accommodate researchers conducting studies in the park, and to highlight how life in the area once was (Papavasileiou, 2008).

The international awards the park has received have turned it into a popular tourist destination for visitors of all nationalities. In addition to individual visitors, organized tours are offered by travel agencies, which transport tourists by bus to the Omalos Plateau. Tourists then hike through the gorge, spend some time in Agia Roumeli for food and rest, and later take a boat to Sfakia, Sougia, or Paleochora, where buses await to pick them up. This serves as an example of the economic network centered around the National Park. Visitor numbers have been systematically recorded since 1981. In 1989, an entrance fee was introduced, initially set at €0.58 (200 drachmas) per person. Over the years, the fee has increased to its current level of €5. The fee is collected at both entrances to the gorge, with most visitors opting for the northern entrance at Xyloskalo.

1.2 Purpose of the Study

Many local businesses, including travel agencies, transportation providers, taverns, restaurants, and hotels, depend on tourism linked to protected areas to boost their revenue. Therefore, the effective management of protected areas is essential to ensuring visitor satisfaction.

This study aims to present innovative insights and significantly contribute to the scientific understanding of visitor satisfaction in Greece's protected areas, focusing on the Samaria National Park.

The study's innovation lies in several aspects:

1. **Addressing the lack of previous studies on protected areas in Greece**
   To this date, there is a significant gap in research on visitor satisfaction in Greece's protected areas. Although Greece hosts numerous areas of exceptional natural beauty and ecological significance, most satisfaction studies have focused on other countries or on other tourism sectors. This study is one of the few to explore visitor experiences in a Greek national park, offering new insights into the management of protected areas.
2. **The unique features and characteristics of the Samaria National Park**
   The Samaria National Park is a unique destination with distinctive characteristics that set it apart from other protected areas in Greece and abroad. Its gorge's geomorphology, rich biodiversity, and cultural significance make it one of the

most popular natural attractions in the country, drawing hundreds of thousands of visitors annually. High visitor numbers in a strictly protected environment present challenges in managing crowds, protecting the natural environment, and ensuring visitor satisfaction. The study analyzes how specific factors influence the visitor experience and highlights how park managers can improve tourism sustainability.

3. **Data from a large sample of visitors over a long period**
   This study is based on over 2000 questionnaires collected during a five-year period (2018–2022), making it one of the few studies with such a large dataset for a protected area. The extended data collection period allows for analyzing changes in visitor satisfaction, monitoring their preferences, and evaluating how external factors, such as the COVID-19 pandemic, influence these preferences. The method that was followed and the findings can be utilized for the management of similar areas.
4. **Comparing conditions before and after the COVID-19 pandemic**
   The COVID-19 pandemic was an exogenous event that dramatically altered the tourism industry and influenced visitor preferences and priorities worldwide. This study is among the few to evaluate visitor experiences and satisfaction factors before and after the pandemic. Analyzing these differences sheds light on behavioral changes and preferences influenced by significant external factors, such as safety perceptions and preferences for less crowded, natural destinations. The analysis of these differences is highly valuable as it highlights changes in visitor behavior and preferences that may be triggered by a significant external factor, new perceptions of safety, and a preference for less crowded, natural destinations. This comparison provides essential insights for the management of protected areas under crisis conditions.
5. **Practical contributions to protected area management**
   The study provides valuable insights and practical recommendations for the management of **Samaria National Park**, which can be utilized by both park managers and other stakeholders involved in the management of protected areas. By using data and analysis from a specific environment, the study offers an analytical approach that can be adapted to other highly visited protected areas.
6. **Enhancing scientific knowledge on sustainable tourism development**
   This study contributes to the broader scientific discussion on sustainable tourism development in protected areas. The research findings can be used to develop strategies that ensure tourism growth does not burden the natural environment but instead promotes a harmonious coexistence between tourism and nature. Data analysis, combined with the proposed recommendations, can contribute to the creation of a sustainable tourism development model based on visitor satisfaction and the protection of natural resources.

The study's goals were:

- To identify the characteristics of Samaria National Park visitors.
- To analyze visitor satisfaction and the factors influencing it.

- To examine differences between the pre- and post-COVID-19 periods.
- To present recommendations for park management and other stakeholders.

Data from the 2018–2022 tourist seasons were used for this study. To collect these data, a questionnaire with well-structured questions was developed in collaboration with the Samaria National Park administrators.

1.3 Structure of the Study

The research is divided into six chapters:

1. **Introduction**
   This chapter includes general information about the study, its structure, and a brief presentation of its objectives, along with key details about the main area of interest, the Samaria National Park.
2. **Literature Review**
   This chapter provides a review of studies related to protected areas worldwide.
3. **Methodology**
   This chapter outlines the methodology used in the study.
4. **Data**
   This chapter presents the data collected for the research.
5. **Results**
   This chapter discusses the analysis and findings.
6. **Discussion and Future Recommendations**
   The final chapter reflects on the results and provides suggestions for future management.

Reference

Papavasileiou, E. (Ed.). (2008). *THE GORGE OF SAMARIA, shelter for life, den of freedom*. Chania.

Chapter 2
Review of the Available Literature

2.1 Tourism as a Global "Industry"

Tourism is one of the largest and fastest-growing industries worldwide (Baloch et al., 2022). Within the tourism industry, Goeldner et al. (2000) identify three main sectors of activity: (1) transportation, (2) hospitality, and (3) attractions. Gunn (1988) argues that attractions are the primary drivers and catalysts for tourism in a specific region. Similarly, Pigram (1983) posits that tourism would not exist without attractions. Attractions play a central role in the tourism industry, and as Gunn (1988) emphasizes, the demand for transportation, accommodation, and related services stemming from tourism would not exist without relevant attractions. Swarbrooke (2000) describes attractions as the lifeblood of the tourism industry, providing motivation for visitors to pursue travel. Similarly, Richards (2001) suggests that although attractions may not be the sole motivation for visitors, they are a focal point for a significant portion of tourism activities and offer a competitive advantage for destinations that have them over other competing destinations. Considering these perspectives, we highlight the importance of attractions for tourism and the need to understand the variety and extent of attractions available in each area.

Numerous efforts have been made to explain the diversity of elements that can be considered attractions (Holloway, 1995; Smith, 1998). According to Buckley et al. (2003), attractions that draw tourists can vary widely, ranging from very small to oversized, accessible for free or requiring a high cost to be accessed, and encompassing natural, human-made, or hybrid features (Prideaux, 2002). Laarman and Durst (1987) define Nature-Based Tourism as a form of tourism characterized by a combination of education, recreation, and adventure. According to Valentine (1992), Nature-Based Tourism primarily involves the direct enjoyment of relatively

C. Zopounidis, G. Fakotakis, *Multi-Criteria Methodology for the Dynamic Development of Tourist Areas*, Multiple Criteria Decision Making,
https://doi.org/10.1007/978-3-031-95314-9_2

undisturbed natural features and includes three categories of activities: those directly connected to nature, those enhanced by nature, and those where nature is a "bonus" element. Jenkins and Pigram (2003) state in their study that activities directly connected to nature include a wide range of recreational actions depending on the natural environment, from relaxing under a tree to exploring the wilderness. Due to the variety of possible nature-based activities, Nature-Based Tourism can take various forms. Research by Newsome et al. (2002) shows that Nature-Based Tourism can take the form of adventure and wildlife encounters or ecotourism, while McKercher (1998) highlights that Nature-Based Tourism includes various forms of non-mass tourism, such as alternative, responsible, and sustainable tourism centered on nature.

2.2 Protected Areas and National Parks

The concept of protected areas has existed since ancient times. Historical evidence shows that across the ancient world, from Greece to India, natural areas were protected and specially managed. A typical example of a protected area in the ancient world is the concept of the "sacred" forest, common in many different cultures. These forests were protected by law in order to be allowed to develop without human intervention (Scoullos et al., 2013).

In modern history, the first protected area was established in 1832 in the United States, when President Andrew Jackson passed legislation to preserve an area around the hot springs of Arkansas. The world's first official National Park was Yellowstone National Park, established 40 years later, in 1872 (Eagles & McCool, 2002). In the years that followed, especially in the twentieth century, many organizations and governments worldwide took action to legally protect various areas of cultural, historical significance, and natural beauty and preserve them for future generations. Various categories of protected areas have been created (WCPA, 1998). In 1948, the International Union for Conservation of Nature (IUCN) was founded by 18 governments, seven international organizations, and 107 national organizations with the ultimate goal of promoting sustainable development and nature conservation. The IUCN defines a protected area as a "clearly defined geographical space, recognized, dedicated, and managed, through legal or other effective means, to achieve the long-term conservation of nature with associated ecosystem services and cultural values" (WCPA, 1998).

The initial goal of protected areas was their absolute protection and preservation, prohibiting any human activity within their boundaries. However, this created problems for the people living around these areas, whose activities and lives were closely tied to them for centuries. Consequently, a redefinition of the initial goal was inevitable, leading to the modern perception that a protected area is closely linked to the surrounding areas on ecological, economic, political, and cultural levels, rather than being an isolated entity (Plummer & Fennell, 2009). Protected areas are no longer accessible solely for scientific research but are associated with a wide range of human activities, and authorities promote the active participation of the local population in the sustainable management of natural resources and cultural values (Scoullos et al., 2013).

National Parks around the world are among the most popular natural attractions (Wearing et al., 2009; Stoleriu et al., 2019). Interestingly, the decision to designate these areas for protection also brings potential challenges arising from tourism (Eagles et al., 2002; Butler, 2018; Niezgoda & Markiewicz, 2021). The designation of National Park status is generally considered an honor for an area and generates the desire for people to visit it (Fredman, 2004; Weiler & Seidl, 2004; Ogden-Jones, 2023). The use of National Parks for tourism purposes is on a continuous rise, driven not only by people's desire to connect with nature or seek related experiences but also by the need for relaxation and engagement in outdoor activities (Davenport & Borrie, 2005; Niezgoda & Nowacki, 2020).

2.3 National Parks in Foreign Countries

2.3.1 National Parks in Germany

Germany is home to 16 designated national parks (Job et al., 2021), mainly located in rural areas, alongside various other protected areas such as biosphere reserves and nature parks. According to German legislation, the primary legal goal for these national parks is the preservation of their ecological integrity (Geyer et al., 2017). However, within these parks, a range of activities is permitted as long as the primary goal of ecological conservation is upheld. These activities include scientific research and guided tours for the public, in line with Germany's Federal Nature Conservation Act.

Researchers in Germany have recognized the potential of tourism in these protected areas and its positive impact on the local economy. The economic benefits generated by tourism in national parks can serve as compensation for local populations, who may face legal restrictions on their activities, thus providing a comparative advantage for the region. Moreover, leveraging and capitalizing on this comparative advantage can lead to additional economic development beyond what would be feasible without the tourism attracted by the national park.

In 2010, a study was published examining the economic impact of tourism in six German national parks: the Lower Saxony Wadden Sea National Park, Bavarian Forest National Park, Eifel National Park, Müritz National Park, Hainich National Park in Thuringia, and Kellerwald-Edersee National Park (Mayer et al., 2010). The study applied the Keynesian multiplier approach, explaining why approaches used for national parks in other countries were less suitable or impractical for Germany. The research relied on both secondary data from the Deutsches Wirtschaftswissenschaftliches Institut für Fremdenverkehr and state statistical agencies, as well as primary data from 78,064 on-site interviews conducted with visitors to the national parks.

Data analysis yielded compelling results. The study not only defined visitor profiles for each park but also identified the types of visitor expenditures. This enabled researchers to quantify the total visitor spending at the parks and, by understanding its structure, calculate both the direct and indirect total economic impact of tourism on local communities. The study also recognized significant differences in visitor numbers, visitor density, and visitor characteristics among different parks, attributed to the unique natural features of each park.

The researchers concluded by acknowledging the limits of the daily visitor capacity that a park can accommodate (carrying capacity) without jeopardizing its protected ecological characteristics. This limitation inevitably places restrictions on the quantitative growth of tourism. For this reason, the study proposed qualitative enhancements, such as increasing overnight stays in local accommodations and offering higher-quality services. This approach would boost per-visitor spending without exerting additional pressure on the natural environment. Another suggestion was the implementation of policies promoting the use of local products by businesses near the parks, thereby increasing the economic multiplier effect of tourist spending. Finally, the study recommended improved marketing and advertising strategies to extend visitors' stay in the broader park area, thereby enhancing the revenue generated for local economies through park-related tourism (Mayer et al., 2010).

2.3.2 National Parks in the United States

National parks in the United States have long been recognized as significant sites, established to protect nature and preserve cultural heritage. They began to be studied as early as the late nineteenth century (Vukomanovic & Randall, 2021). In 1916, the U.S. government established the National Park Service (NPS) to manage and oversee all national parks, national monuments, and historically significant areas at the federal level. Prior to this, local management of each site was the norm. The NPS is responsible for managing 401 sites across the United States, encompassing more than 84 million acres of land. Its primary objectives are the protection and preservation of these areas' unique natural beauty while providing access to the general public (United States Department of the Interior, 2012).

Protected areas, particularly national parks in the United States, are renowned for the vast number of tourists they attract annually. Recent data indicates they draw more than 3.3 billion visitors annually, underscoring their importance as premier tourist destinations (Balmford et al., 2015). Policymakers and researchers have long recognized the economic significance of national parks, considering them valuable assets for the surrounding areas and the national economy. For over 30 years, the NPS has meticulously monitored visits to each site it manages. Researchers collect data on visitor numbers and average spending through surveys, as well as local and national economic multipliers (United States Department of the Interior, 2012).

From 1997 to 2011, the second edition of the *Money Generation Model (MGM2)*, developed by *Daniel Stynes* and Dennis Propst based on the original *Money Generation Model* by *Ken Hornback,* was used to estimate the total economic impact of expenditures related to visits to *National Parks*. This economic impact was calculated annually for the areas surrounding the parks, the state in which the park is located, and the country as a whole (United States Department of the Interior, 2012).

Starting in *2012*, the *MGM2 model* was replaced by the *Visitor's Spending Effects (VSE) model*, a collaborative effort between the *NPS (National Park Service)* and the *U.S. Geological Survey*. While the *VSE model* retained the core processes of *MGM2*, it significantly improved the accuracy of economic impact analysis.

Both models use *economic multipliers* derived from the *IMPLAN input-output modeling system*. However, the *VSE model* introduced *park-specific multipliers* based on *county-level economic data*, making estimates more precise. Additionally, *gateway regions* around each park, typically within a *60-mile radius*, were redefined using *GIS data*, allowing the *VSE model* to capture a larger share of *secondary spending* and provide *more accurate local estimates*.

Furthermore, the *VSE model* created *visitor profiles* based on *VSP (Visitor Spending Profile) data*, categorizing parks into four types according to *accommodation availability and visitor overnight trends*, leading to even more precise results.

Notably, due to these methodological changes, *economic impact results from 2012 onward are not directly comparable* to those of previous years (United States Department of the Interior, 2012).

The economy is characterized by a complex system of interactions between producers and consumers within the same and across different economic sectors. The goods produced by businesses operating in one sector can serve as final products for consumers or as inputs for other sectors, creating a network of interdependencies (Colacchio & Vergori, 2023). This complexity also applies to tourism in national parks, making the calculation of its overall economic impacts intricate. Visitors spend money on various goods and services in the areas surrounding National Parks, generating sales, income, and employment for local businesses (direct economic effect). To produce their products, local businesses require inputs sourced from other sectors, creating additional economic activity (indirect economic effect). Furthermore, employees in businesses directly and indirectly affected spend their earnings, contributing to the local economy (induced effect). The sum of the indirect and induced effects constitutes the secondary economic effect. Adding this to the direct economic effect yields the total economic impact. The VSE model, used by the NPS (as well as the MGM2 before it), aims to capture all these effects to provide an accurate description of the overall economic impact of visiting a National Park on the economy (United States Department of the Interior, 2012).

The economic impact can be categorized into two main types: "economic contributions" and "economic impacts." Economic contributions represent the relative significance of the total economic activity generated by visitor spending in National Parks for local economies. This is determined by multiplying the total visitor expenditures by regional economic multipliers. In contrast, economic impacts refer to the economic consequences of spending by visitors coming from other locations, who would likely not have traveled to the area if not for the presence of the specific National Park.

In both of these categories, four distinct effects can be identified:

Output: This represents the total value of production associated with businesses linked to visits to National Parks, encompassing both transactions between businesses and final demand from consumers.

Employment: This indicates the jobs created as a result of economic activity related to visits to a National Park.

Labor Income: This includes wages and salaries paid to employees working in businesses associated with visits to a National Park.

Value Added: This represents the economic value above the cost of production for a service or product based on its selling price (United States Department of the Interior, 2012).

Taking all of the above into account, it is widely accepted that the United States National Park System constitutes a valuable economic asset. It has been recognized that it generates more than four dollars in value for every dollar allocated to its annual budget by the government. Furthermore, the economic growth associated with businesses in counties surrounding National Parks tends to surpass the growth observed in areas without parks (United States National Park Conservation Association, 2006). More specifically, in 2019, U.S. national parks contributed over $41 billion to the economy and created more than 340,000 jobs (U.S. Department of the Interior, 2020).

2.3.3 National Parks in Canada

Canada is also renowned for its vast natural beauty, with protected areas covering approximately 9.5% of its total land area. The Parks Canada Agency, established in 1911, oversees the management of national parks, national historic sites, and national marine protected areas. These parks consistently attract millions of visitors annually, contributing billions of dollars to the economy through expenditures on goods and services (Parks Canada Agency, 2011).

The discussion surrounding a comprehensive national analysis of the economic impacts of these protected areas began in Canada in 2003. Efforts to collect data covering multiple locations culminated in the publication of the first related report in 2005. This initial report estimated the total visitor expenditures at $2.98 billion compared to the total expenditures of $554.6 million by park management authorities. It also focused on the broader national economic impacts resulting from visitor spending, including a $2.456 billion contribution to the Gross Domestic Product (GDP), $1.634 billion in labor income, the creation of 59,021 full-time jobs, and $137 million in tax revenues.

However, as this was the first attempt at such a comprehensive study in Canada, challenges arose in obtaining consistent and reliable data. The insights gained from this initial effort laid the groundwork for subsequent studies, enabling local authorities to refine their data collection methods and achieve more accurate and reliable results in later analyses (Canadian Parks Council, 2011).

2.3.4 *National Parks in England*

In England, the legal entity known as a "National Park" was legislated in 1949 with the National Parks and Access to the Countryside Act. This legislation recognized the need to preserve and protect specific areas distinguished by their exceptional natural beauty, rich wildlife, and cultural heritage. The primary goal was to encourage the public to enjoy and appreciate these areas. British policymakers regarded national parks as national treasures with significant economic value. These parks cover a total area of more than 1.2 million hectares, accounting for approximately 9.3% of England's total land area. In 1995, the Environment Act established National Park Authorities tasked with creating comprehensive Management Plans for the parks and implementing them in close collaboration with local communities, businesses operating in the respective areas, and other local organizations and stakeholders (National Parks England, 2013).

England's national parks attract a substantial number of visitors annually, making them key destinations for tourism and recreation. Hotels, restaurants, and various tourism-related businesses are well-established in the local economies surrounding the parks. To evaluate the economic value of tourism in these parks, the National Park Authorities have conducted studies using the Scarborough Tourism Economic Activity Monitor (STEAM).

The findings reveal an annual influx of 95 million visitors, of which 87 million are day-trippers, and 24 million are overnight visitors. Total annual tourist spending amounts to £3 billion. Naturally, these figures vary significantly among parks, depending on the overall number of visitors to each park, particularly the number of overnight visitors, who tend to spend significantly more than day-trippers (National Parks England, 2013). The total visitor spending helps sustain 48,000 full-time jobs within the boundaries of the national parks, as well as a conservatively estimated 20,000 full-time jobs in the broader areas surrounding the parks (National Parks England, 2013).

2.4 The Contribution of Nature-Based Tourism to the Green Economy

It is widely acknowledged that certain segments of the tourism sector can be considered a "clean industry" and part of the Green Economy. Specifically, nature-based tourism is a significant sector within this industry (Balmford et al., 2009; Dedeke,

2017). This form of tourism caters to the growing consumer demand for reconnecting with nature (Stoleriu et al., 2019; Niezgoda & Nowacki, 2020). This demand stems from the fact that nature generally promotes well-being, both physically and psychologically (Kovacs-Györi et al., 2018; Schwartz et al., 2019; Plunz et al., 2019; Zhang et al., 2019). Additionally, nature serves as a sanctuary for biodiversity, allowing people to enjoy and rejuvenate within its tranquil environment (Kaffashi et al., 2015; Schwartz et al., 2019).

Protecting areas with significant biodiversity and landscapes of outstanding natural beauty is a core pillar of nature-based tourism. Protected areas and nature-based tourism offer visitors access to these ecosystems (MEA, 2005; Hausmann et al., 2016). National parks, in particular, are characterized by high levels of biodiversity protection compared to other categories of protected areas. Furthermore, they facilitate structured tourism by organizing local economies around them (Mangachena & Pickering, 2021; Vurnek et al., 2018). As a result, national parks play a vital role in the tourism sector.

It is essential to analyze the factors that attract visitors and make visits to protected areas enjoyable. Internal factors, such as expectations regarding destinations and activities, and external factors, including accessibility and transportation options, significantly influence tourists' perceptions of natural landscapes (Stoleriu et al., 2019). Consequently, businesses providing nature-based tourism services and related stakeholders must not ignore the diverse opinions visitors hold about nature and recreational activities. Understanding how people perceive their recreational experiences in these protected areas has become fundamental (Hausmann et al., 2020).

2.5 Visitor Satisfaction and National Parks

The study of customer satisfaction is foundational in consumer behavior research due to its critical role in predicting future behaviors and outcomes (Huang et al., 2015; Malter et al., 2020; Choi et al., 2020).

Customer satisfaction relates to the extent to which a product or service meets or exceeds expectations (Kotler & Armstrong, 2018). In tourism, satisfaction evaluates the degree to which a destination fulfills tourists' expectations during their trip. It is widely accepted that customer satisfaction ensures a business demonstrates value to its clients through its offered services or products (Zameer et al., 2015).

Satisfied customers generally remain loyal to products and services that effectively address their needs and desires (Mittal et al., 2023). In tourism, satisfied tourists are more likely to return to destinations, which, in turn, enhances the location's future performance in the tourism sector. Satisfaction is particularly crucial in tourism, as travelers often invest significant financial resources in their trips. They want assurances that their experience will be worth the expense, meeting or exceeding their expectations.

Evaluating visitor satisfaction with services and the overall quality of experiences in national parks is paramount for park management and local tourism stakeholders (Eagles, 2001; Tonge et al., 2011; Arnberger et al., 2012; Balmford et al., 2015). Visitor satisfaction significantly impacts local economic development and the sustainable management of protected areas. A positive experience can lead to repeat visits, increased spending, recommendations to friends and family, and the attraction of new tourists (Lee et al., 2007; Rivera & Croes, 2010; Taplin et al., 2016; Moore et al., 2017).

Achieving high levels of visitor satisfaction should be a priority for nature-based tourism destinations (Moore et al., 2015). Satisfaction is a critical emotion in tourist behavior research, providing insight into their enjoyment, needs, desires, and willingness to spend (Pearce, 2005). It plays a pivotal role in the success and future of tourism-related products and services (Gursoy et al., 2007).

Most researchers in the 1980s and 1990s concluded that satisfaction is more of an evaluative judgment rather than merely an attitude. It is a complex subject with numerous related measurement issues (Yüksel & Rimmington, 1998). Satisfaction can be measured holistically for a product, service, or destination or at the level of individual attributes contributing to satisfaction. Attribute satisfaction is defined as the consumer's subjective judgment regarding the performance of features (Oliver, 1993). Satisfaction with specific attributes can then be aggregated to calculate overall satisfaction.

It is essential to understand the significance of each feature to customers. Studies have identified that economic, ecological, and social sustainability are vital factors for satisfaction in nature-based tourism (Atsbha & Kumar, 2019). Other studies include price, perceived value, and responsibility (Akama & Kieti, 2003), as well as communication and empathy (Frochot, 2003). Physical characteristics play a more prominent role in visitor satisfaction in national parks (Mehmetoglu, 2007). The primary satisfaction factor remains the pristine environment (Prato, 2006), although other factors are also considered significant.

Tourism satisfaction research has traditionally relied on models such as the Expectation-Disconfirmation Theory (Oliver, 1980) and the Perceived Performance Model (Tse & Wilton, 1988). The Expectation-Disconfirmation Model posits that satisfaction is related to the relationship between perceptions and customer expectations. Satisfaction results from an evaluative reaction to a purchase or consumption experience and the comparison between expectations and actual performance. If performance exceeds expectations, customers are likely to have a positive reaction, leading to high satisfaction and an intention to repurchase the product.

Based on this theory, the SERVQUAL tool was developed to measure perceived service quality (Parasuraman et al., 1988). The Perceived Performance Model (Tse & Wilton, 1988) suggests that customer satisfaction with a service or product can be assessed solely by examining actual performance evaluation. This model implies that customers are likely satisfied when a product or service operates at a desirable level, regardless of prior expectations.

Since service quality is under the control of the provider of a product or service, this model may be more useful than other visitor satisfaction theories as a basis for developing tools for tourism providers to achieve an accurate assessment of visitor satisfaction (Baker & Crompton, 2000).

It is crucial to recognize and measure visitor satisfaction for every feature of a destination, as satisfaction or dissatisfaction with one element may influence overall satisfaction with the destination (Kozak & Rimmington, 2000). Achieved satisfaction levels significantly affect future intentions, such as revisiting a destination and recommending it to others. Customer loyalty reflects attitudes and behaviors toward services and the intention to repeatedly use them (Backman & Crompton, 1991).

Studies have shown that understanding factors contributing to increased loyalty is valuable for tourism marketing and management professionals (Flavian et al., 2001). Future customer behavior can be measured using two indicators: repeat buy intention and willingness to recommend (Cronin & Taylor, 1992).

In tourism research, a similar approach is adopted, where destination loyalty is expressed in terms of the intention to revisit in the future and the willingness to recommend the destination to friends and family (Oppermann, 2000). Therefore, two indicators are used to measure loyalty: "intention to revisit" and "intention to recommend."

In nature-based tourism, recommendations to others may serve as a more useful indicator of loyalty compared to future visits. This is because visitors may seek new experiences even if they were satisfied with their previous destination (Chen & Gursoy, 2001).

2.6 The Impact of COVID-19

The COVID-19 pandemic created significant challenges across public health, the economy, society, and the environment (WHO, 2020; Waithaka et al., 2021). It led to the closure or restricted operation of many national parks worldwide, including the suspension of social gatherings and events, particularly during the initial stages of the pandemic (Hockings et al., 2020; Waithaka et al., 2021). Measures such as quarantines and travel restrictions significantly affected park visitation, tourism revenues, park management operations, and local stakeholders (Hockings et al., 2020; Kupfer et al., 2021; Smith et al., 2021; Spenceley et al., 2021; Mandić, 2021).

The pandemic's impacts on nature-based tourism varied across regions, providing an opportunity to explore different park management strategies (Waithaka et al., 2021; Kupfer et al., 2021). Measures against the spread COVID-19, such as restrictions on social interaction and mobility, stay-at-home orders, and event cancellations, intensified public demand for access to natural spaces and outdoor recreational activities (Geng et al., 2021). These changes highlighted the critical and unique role nature plays in supporting physical and psychological health.

The Samaria National Park was also significantly affected, as the 2020 tourist season's opening was delayed until the necessary operational mechanisms for the new conditions were established. Nevertheless, visitor interest remained strong despite the prevailing circumstances.

Within the context of the COVID-19 pandemic, visitor experiences in national parks became increasingly significant for several reasons. Firstly, the pandemic underscored the heavy reliance of local communities around national parks on tourism, such as the village of Agia Roumeli, which primarily serves visitors who traverse the Samaria National Park. Thus, effective management that ensures visitor safety alongside enjoyable experiences in these parks contributes substantially to improving local business revenues and promoting sustainable tourism.

Additionally, lifestyle changes induced by public health directives during the pandemic presented various challenges. Health crises lead to economic uncertainty and safety concerns, which significantly impacted individuals both physically and psychologically (Wu et al., 2005; Monson et al., 2017; Brooks et al., 2020; Gao et al., 2020; Xiang et al., 2020; Bo et al., 2020; Geng et al. 2021, 2021). Within this context, national parks offered a vital refuge, providing expansive, safe areas for outdoor activities (Seaman et al., 2010; Annerstedt et al., 2012). This not only mitigated the adverse effects of the pandemic but also allowed visitors to enjoy the natural environment (Hockings et al., 2020; Nicola et al., 2020).

Effective management of safety and visitor experiences in national parks was essential in reducing the pandemic's impact on social well-being, physical health, economic prosperity, and tourism. Consequently, national parks played a pivotal role in addressing the challenges of the COVID-19 pandemic.

As parks returned to full operational capacity, a notable increase in visitation was observed (Templeton et al., 2021). This makes it imperative for park managers to understand the implications of large visitor numbers on both visitor satisfaction and tourism safety. Ensuring that visitors have safe and enjoyable experiences post-pandemic is critical to maintaining high satisfaction levels.

Tourism remains vulnerable to potential future health crises and the socio-economic consequences of the COVID-19 pandemic (Spenceley et al., 2021). Improving visitor experiences in national parks and striving for sustainable tourism—both now and in the future—is of utmost importance. To achieve this, it is vital to understand any changes in visitor satisfaction levels and the factors determining satisfaction. Additionally, shifts in visitor demographics, perceptions, and visitation frequency before and after COVID-19 should be documented.

References

Akama, J. S., & Kieti, D. M. (2003). Measuring tourist satisfaction with Kenya's safari: A case study of Tsavo West National Park. *Tourism Management, 2*, 73–81.

Annerstedt, M., Östergren, P. O., Björk, J., Grahn, P., Skärbäck, E., & Währborg, P. (2012). Green qualities in the neighbourhood and mental health-results from a longitudinal cohort study in Southern Sweden. *BMC Public Health, 12*(1), 1–13.

Arnberger, A., Eder, R., Allex, B., Sterl, P., & Burns, R. C. (2012). Relationships between national-park affinity and attitudes towards protected area management of visitors to the Gesaeuse National Park, Austria. *Forest Policy and Economics, 19*, 48–55.

Atsbha, G. A., & Kumar, S. (2019). The structural relationship between tourist satisfaction and sustainable heritage tourism development in Tigrai, Ethiopia. *Heliyon, 5*(3), e01335.

Backman, S., & Crompton, J. (1991). The usefulness of selected variables for predicting activity loyalty. *Leisure Sciences, 13*, 205–220.

Baker, D., & Crompton, J. (2000). Quality, satisfaction and behavioral intentions. *Annals of Tourism Research., 27*, 785–804.
Balmford, A., Beresford, J., Green, J., Naidoo, R., Walpole, M., & Manica, A. (2009). A global perspective on trends in nature-based tourism. *PLoS Biology, 7*, e1000144.
Balmford, A., Green, J., Anderson, M., Beresford, J., Huang, C., Naidoo, R., & Walpole, M. (2015). Walk on the wild side: Estimating the global magnitude of visits to protected areas. *PLoS Biology, 13*(2), e1002074.
Baloch, Q., Shah, S., Iqbal, N., Sheeraz, M., Asadullah, M., Mahar, S., & Khan, A. (2022). Impact of tourism development upon environmental sustainability: A suggested framework for sustainable ecotourism. *Environmental Science and Pollution Research, 30*, 5917.
Bo, H. X., Li, W., Yang, Y., Wang, Y., Zhang, Q., Cheung, T., & Xiang, Y. T. (2020). Posttraumatic stress symptoms and attitude toward crisis mental health services among clinically stable patients with COVID-19 in China. *Psychological Medicine, 51*(6), 1052–1053.
Brooks, S. K., Webster, R. K., Smith, L. E., Woodland, L., Wessely, S., Greenberg, N., & Rubin, G. J. (2020). The psychological impact of quarantine and how to reduce it: Rapid review of the evidence. *The Lancet, 395*(10227), 912–920.
Buckley, R., Pickering, C. M., & Weaver, D. (2003). *Nature-based tourism, environment and land management*. CABI Publishing.
Butler, R. (2018). Sustainable tourism in sensitive environments: A wolf in sheep's clothing? *Sustainability, 10*(6), 1–11.
Canada. Canadian Parks Council. (2011). *The economic impact of Canada's National, Provincial & Territorial Parks in 2009*. The Outspan Group Inc.
Canada. Parks Canada Agency. (2011). *The economic impact of Parks Canada*. The Outspan Group Inc.
Chen, J., & Gursoy, D. (2001). An investigation of tourists' destination loyalty and preferences. *International Journal of Contemporary Hospitality Management., 13*, 79–85.
Choi, H. C., Huang, S., Choi, H., & Chang, H. S. (2020). The effect of flight attendants' physical attractiveness on satisfaction, positive emotion, perceived value, and behavioral intention. *Journal of Hospitality and Tourism Management, 44*, 19–29.
Colacchio, G., & Vergori, A. S. (2023). Tourism development and Italian economic growth: The weight of the regional economies. *Journal of Risk and Financial Management, 16*(4), 245.
Cronin, J., & Taylor, S. (1992). Measuring service quality: A re-examination and extension. *Journal of Marketing, 56*(3), 55–68.
Davenport, M. A., & Borrie, W. T. (2005). The appropriateness of snowmobiling in National Parks: An investigation of the meanings of snowmobiling experiences in Yellowstone National Park. *Environmental Management, 35*, 151–160.
Dedeke, A. N. (2017). Creating sustainable tourism ventures in protected areas: An actor-network theory analysis. *Tourism Management, 61*, 161–172.
Eagles, P. F. J. (2001). International trends in park tourism. In *Proceedings of the EUROPARC Federation 2001*, Matrei, Austria, 3–7 October 2001.
Eagles, P. F. J., & McCool, S. F. (2002). *Tourism in National Parks and protected areas: Planning and management*. Cabi Publishing.
Eagles, P. F. J., McCool, S. F., & Haynes, C. D. (2002). *Sustainable tourism in protected areas: Guidelines for planning and management*. IUCN Gland.
Flavian, C., Martinez, E., & Polo, Y. (2001). Loyalty to grocery stores in the Spanish market of the 1990s. *Journal of Retailing and Consumer Services, 8*, 85–93.
Fredman, P. (2004). National Park designation: Visitor flows and tourism impact. In *Policies, methods and tools for visitor management: Proceedings of the second international conference on monitoring and management of visitor flows in recreational and protected areas, June 16–20, 2004, Rovaniemi, Finland* (pp. 369–375).
Frochot, I. (2003). An analysis of regional positioning and its associated food images in French tourism regional brochures. *Journal of Travel and Tourism Marketing, 14*, 77–96.

Gao, J., Zheng, P., Jia, Y., Chen, H., Mao, Y., Chen, S., & Dai, J. (2020). Mental health problems and social media exposure during COVID-19 outbreak. *PLoS One, 15*(4), e0231924.

Geng, D. C., Innes, J., Wu, W., & Wang, G. (2021). Impacts of COVID-19 pandemic on urban park visitation: A global analysis. *Journal of Forest Research, 32*(2), 553–567.

Geng, D. C., Innes, J. L., Wu, W., Wang, W., & Wang, G. (2021). Seasonal variation in visitor satisfaction and its management implications in Banff National Park. *Sustainability, 13*(4), 1681.

Geyer, J., Kreft, S., Jeltsch, F., & Ibisch, P. L. (2017). Assessing climate change-robustness of protected area management plans—The case of Germany. *PLoS One, 12*(10), e0185972.

Goeldner, C., Ritchie, J., & McIntosh, R. (2000). *Tourism: Principles, practices, philosophies*. Wiley.

Gunn, C. (1988). *Vacationscape: Designing tourist regions* (2nd ed.). Van Nostrand Reinhold.

Gursoy, D., Mccleary, K., & Lepsito, L. (2007). Propensity to complain: Effects of personality and behavioral factors. *Journal of Hospitality & Tourism Research., 31*, 358–386.

Hausmann, A., Slotow, R., Burns, J. K., & Di Minin, E. (2016). The ecosystem service of sense of place: Benefits for human Well-being and biodiversity conservation. *Environmental Conservation, 43*, 117–127.

Hausmann, A., Toivonen, T., Fink, C., Heikinheimo, V., Kulkarni, R., Tenkanen, H., & Di Minin, E. (2020). Understanding sentiment of national park visitors from social media data. *People and Nature, 2*, 750–760.

Hockings, M., Dudley, N., Ellio, W., Ferreira, M. N., & Yang, A. (2020). COVID-19 and protected and conserved areas. *Parks, 26*(1), 7–24.

Holloway, L. C. (1995). *The business of tourism* (4th ed.). Longman.

Huang, S., Shen, Y., & Choi, C.S. (2015). The effects of motivation, satisfaction and perceived value on tourist recommendation.

Jenkins, J., & Pigram, J. (2003). *Encyclopedia of leisure and outdoor recreation*. Routledge.

Job, H., Bittlingmaier, S., Mayer, M., von Ruschkowski, E., & Woltering, M. (2021). Park–people relationships: The socioeconomic monitoring of national parks in Bavaria. *Germany. Sustainability, 13*(16), 8984.

Kaffashi, S., Radam, A., Shamsudin, M. N., Yacob, M. R., & Nordin, N. H. (2015). Ecological conservation, ecotourism, and sustainable management: The case of Penang National Park. *Forests, 6*, 2345–2370.

Kotler, P., & Armstrong, G. M. (2018). *Principles of marketing*. Pearson.

Kovacs-Györi, A., Ristea, A., Kolcsar, R., Resch, B., Crivellari, A., & Blaschke, T. (2018). Beyond spatial proximity-classifying parks and their visitors in London based on spatiotemporal and sentiment analysis of Twitter data. *ISPRS International Journal of Geo-Information, 7*, 378.

Kozak, M., & Rimmington, M. (2000). Tourist satisfaction with Mallorca, Spain, as of season holiday destination. *Journal of Travel Research, 38*, 260–269.

Kupfer, J. A., Li, Z., Ning, H., & Huang, X. (2021). Using mobile device data to track the effects of the COVID-19 pandemic on spatiotemporal patterns of national park visitation. *Sustainability, 13*(16), 9366.

Laarman, J. G., & Durst, P. B. (1987). Nature travel in the tropics. *Journal of Forestry, 85*(5), 43–46.

Lee, J., Graefe, A. R., & Burns, R. C. (2007). Examining the antecedents of destination loyalty in a forest setting. *Leisure Sciences, 29*, 463–481.

Malter, M. S., Holbrook, M. B., Kahn, B. E., Parker, J. R., & Lehmann, D. R. (2020). The past, present, and future of consumer research. *Marketing Letters, 31*(2–3), 137–149.

Mandić, A. (2021). Protected area management effectiveness and COVID-19: The case of Plitvice Lakes National Park, Croatia. *Journal of Outdoor Recreation and Tourism, 41*, 100397.

Mangachena, J. R., & Pickering, C. M. (2021). Implications of social media discourse for managing national parks in South Africa. *Journal of Environmental Management, 285*, 112159.

Mayer, M., Móller, M., Woltering, M., Arnegger, J., & Job, H. (2010). The economic impact of tourism in six German national parks. *Landscape and Urban Planning, 97*, 73–82.

McKercher, B. (1998). *The business of nature-based tourism*. Hospitality Press.

MEA Millennium Ecosystem Assessment (MA). (2005). *Ecosystems and human well-being: Synthesis*. Island Press.
Mehmetoglu, M. (2007). Typologising nature-based tourists by activity-theoretical and practical implications. *Tourism Management, 28*, 651–660.
Mittal, V., Han, K., Frennea, C., Blut, M., Shaik, M., Bosukonda, N., & Sridhar, S. (2023). Customer satisfaction, loyalty behaviors, and firm financial performance: What 40 years of research tells us. *Marketing Letters, 34*, 1–17. https://doi.org/10.1007/s11002-023-09671-w
Monson, E., Caron, J., McCloskey, K., & Brunet, A. (2017). Longitudinal analysis of quality of life across the trauma spectrum. *Psychological Trauma: Theory, Research, Practice, and Policy, 9*(5), 605–612.
Moore, S. A., Rodger, K., & Taplin, R. (2015). Moving beyond visitor satisfaction to loyalty in nature-based tourism: A review and research agenda. *Current Issues in Tourism, 18*, 667–683.
Moore, S. A., Rodger, K., & Taplin, R. H. (2017). Developing a better understanding of the complexities of visitor loyalty to Karijini National Park, Western Australia. *Tourism Management, 62*, 20–28.
Newsome, D., Moore, S. A., & Dowling, R. K. (2002). *Natural area tourism: Ecology, impacts and management*. Channel View Publications.
Nicola, M., Alsafi, Z., Sohrabi, C., Kerwan, A., Al-Jabir, A., Iosifidis, C., Agha, M., & Agha, R. (2020). The socio-economic implications of the coronavirus pandemic (COVID-19): A review. *International Journal of Surgery, 78*, 185–193. https://doi.org/10.1016/j.ijsu.2020.04.018
Niezgoda, A., & Markiewicz, E. (2021). The effects of the COVID-19 pandemic in National Parks—Diagnosis and forecast of sustainable tourism. *Studia Peregietica, 33*(1), 9–24.
Niezgoda, A., & Nowacki, M. (2020). Experiencing nature: Physical activity, beauty and tension in Tatra National Park-analysis of TripAdvisor reviews. *Sustainability, 12*(2), 1–15.
Ogden-Jones, R. (2023). Narrative and story-telling in Darwall & Anor v Dartmoor National Park authority. *Journal of Environmental Law., 35*, 437.
Oliver, R. L. (1980). A cognitive model of the antecedents and consequences of satisfaction decisions. *Journal of Marketing Research, 17*, 460–469.
Oliver, R. (1993). Cognitive, affective and attribute bases of the satisfaction response. *Journal of Consumer Research, 20*, 418–430.
Oppermann, M. (2000). Tourism destination loyalty. *Journal of Travel Research, 39*, 78–84.
Parasuraman, A., Zeithaml, V. A., & Berry, L. L. (1988). SERVQUAL: A multiple-item scale for measuring consumer perceptions of service quality. *Journal of Retailing, 64*(1), 12–40.
Pearce, P. L. (2005). *Tourist behaviour: Themes and conceptual schemes*. Channel View Publications.
Pigram, J. J. (1983). *Outdoor recreation resource management*. Croom Helm.
Plummer, R., & Fennell, D. A. (2009). Managing protected areas for sustainable tourism: Prospects for adaptive co-management. *Journal of Sustainable Tourism, 17*(2), 149–168.
Plunz, R. A., Zhou, Y., Carrasco Vintimilla, M. I., McKeown, K., Yu, T., Uguccioni, L., & Sutto, M. P. (2019). Twitter sentiment in New York City parks as measure of well-being. *Landscape and Urban Planning, 189*, 235–246.
Prato, T. (2006). *Using multi-criteria design analysis in natural resource management*. Ashgate Publishing.
Prideaux, B. (2002). Building visitor attractions in peripheral areas. *International Journal of Tourism Research, 4*, 379–391.
Richards, G. (2001). *Cultural attractions and European tourism*. CABI Publishing.
Rivera, M. A., & Croes, R. (2010). Ecotourists' loyalty: Will they tell about the destination or will they return? *Journal of Ecotourism, 9*, 85–103.
Schwartz, A. J., Dodds, P. S., O'Neil-Dunne, J. P. M., Danforth, C. M., & Ricketts, T. H. (2019). Visitors to urban greenspace have higher sentiment and lower negativity on Twitter. *People and Nature, 1*, 476–485.

Scoullos, M., et al. (2013). *Education for sustainable development in biosphere reserves and other designated areas: A resource book for educators in South-Eastern Europe and the Mediterranean*. UNESCO.

Seaman, P. J., Jones, R., & Ellaway, A. (2010). It's not just about the park, it's about integration too: Why people choose to use or not use Urban greenspaces. *International Journal of Behavioral Nutrition and Physical Activity, 7*(1), 1–9.

Smith, R. (1998). Visitors attractions in Scotland. In R. McLellan & R. Smith (Eds.), *Tourism in Scotland* (pp. 187–208). International Thompson Business Press.

Smith, M. K. S., Smit, I. P., Swemmer, L. K., Mokhatla, M. M., Freitag, S., Roux, D. J., & Dziba, L. (2021). Sustainability of protected areas: Vulnerabilities and opportunities as revealed by COVID-19 in a national park management agency. *Biological Conservation, 255*, 108985.

Spenceley, A., McCool, S., Newsome, D., Báez, A., Barborak, J. R., Blye, C. J., Bricker, K., Cahyadi, H. S., Corrigan, K., Halpenny, E., Hvenegaard, G. T., King, D. M., Leung, Y. F., Mandić, A., Naidoo, R., Rueede, D., Sano, J. A., Sarhan, M., Santamaria, V., & Zschiegner, A. K. (2021). Tourism in protected and conserved areas amid the COVID-19 pandemic. *Parks, 27*, 103–118.

Stoleriu, O. M., Brochado, A., Rusu, A., & Lupu, C. (2019). Analyses of visitors' experiences in a natural world heritage site based on TripAdvisor reviews. *Visitor Studies, 22*(2), 192–212.

Swarbrooke, J. (2000). *Sustainable tourism management*. CABI International.

Taplin, R. H., Rodger, K., & Moore, S. A. (2016). A method for testing the effect of management interventions on the satisfaction and loyalty of national park visitors. *Leisure Sciences, 38*, 140–160.

Task Force on Economic Benefits of Protected Areas of the World Commission on Protected Areas (WCPA) of IUCN, in collaboration with the Economics Service Unit of IUCN. (1998). *Economic values of protected areas: Guidelines for protected area managers*. IUCN.

Templeton, A. J., Goonan, K., & Fyall, A. (2021). COVID-19 and its impact on visitation and management at US national parks. *International Hospitality Review, 35*(2), 240–259.

Tonge, J., Moore, S., & Taplin, R. (2011). Visitor satisfaction analysis as a tool for park managers: A review and case study. *Annals of Leisure Research., 14*, 289. https://doi.org/10.1080/11745398.2011.639339

Tse, D., & Wilton, P. (1988). Models of consumer satisfaction formation: An extension. *Journal of Marketing Research, 25*, 204–212.

U.S. Department of the Interior. (2020). National park visitor spending contributed $40 Billion to U.S. Economy | U.S. Department of the Interior. Available from: https://www.doi.gov/pressreleases/national-park-visitor-spending-contributed-40-billion-us-economy

United Kingdom. National Parks England. (2013). *Valuing England's National Parks*. Comulus Consultants Ltd and ICF GHK.

United States. Department of the Interior. (2012). *BUDGET JUSTIFICATIONS and performance information fiscal year 2012*. National Park Service.

United States. National Park Conservation Association. (2006). *The U.S. National Park System: An economic asset at risk*. Hardner & Gullison Associates, LLC. Available from: http://www.npca.org/assets/pdf/NPCA_Economic_Significance_Report.pdf

Valentine, P. S. (1992). Review: Nature-based tourism. In B. Weiler & C. M. Hall (Eds.), *Special interest tourism* (pp. 105–127). Belhaven.

Vukomanovic, J., & Randall, J. (2021). Research trends in U.S. national parks, the world's "living laboratories". *Conservation Science and Practice, 3*, e414.

Vurnek, M., Brozinčević, A., Culinović, K., & Novosel, A. (2018). Challenges in the management of Plitvice Lakes National Park, Republic of Croatia. In M. N. Suratman (Ed.), *National Parks—Management and conservation* (pp. 55–72). IntechOpen.

Waithaka, J., Dudley, N., Álvarez, M., Mora, S. A., Chapman, S., Figgis, P., Fitzsimons, J., Gallon, S., Gray, T. N. E., Kim, M., Pasha, M. K. S., Perkin, S., Boixeda, P. R., Sierra, C., Valverde, A., & Wong, M. (2021). Impacts of COVID-19 on protected and conserved areas: A global overview and regional perspectives. *Parks, 27*, 85. https://doi.org/10.2305/IUCN.CH.2021.PARKS-27-SIJW.en

Wearing, S., Scheinsberg, S., Grabowski, S., & Tumes, K. (2009). *Understanding track/trail experiences in National Parks. A review*. CRC for Sustainable Tourism.
Weiler, S., & Seidl, A. (2004). What's in a name? Extracting econometric drivers to assess the impact of national park designation. *Journal of Regional Science, 44*(2), 245–262.
WHO (2020). WHO Director-General's opening remarks at the media briefing on COVID-19 - 11 March 2020. Retrieved from https://www.who.int/director-general/speeches/detail/who-director-general-s-opening-remarks-at-the-media-briefing-on-covid-19%2D%2D-11-march-2020
Wu, K. K., Chan, S. K., & Ma, T. M. (2005). Posttraumatic stress, anxiety, and depression in survivors of severe acute respiratory syndrome (SARS). *Journal of Traumatic Stress, 18*(1), 39–42.
Xiang, Y. T., Yang, Y., Li, W., Zhang, L., Zhang, Q., Cheung, T., & Ng, C. H. (2020). Timely mental health care for the 2019 novel coronavirus outbreak is urgently needed. *The Lancet Psychiatry, 7*(3), 228–229.
Yüksel, A., & Rimmington, M. (1998). Customer-satisfaction measurement. *Cornell Hotel and Restaurant Administration Quarterly, 39*(6), 60–71.
Zameer, H., Tara, A., Kausar, U., & Mohsin, A. (2015). Impact of service quality, corporate image and customer satisfaction towards customers' perceived value in the banking sector in Pakistan. *International Journal of Bank Marketing, 33*(4), 442–456.
Zhang, T., Zhang, W., Meng, H., & Zhang, Z. (2019). Analyzing visitors' preferences and evaluation of satisfaction based on different attributes, with forest trails in the Akasawa National Recreational Forest, Central Japan. *Forests, 10*, 431.

Chapter 3
Methodology

3.1 Multi-Criteria Analysis

Multi-Criteria Analysis (MCA) is a decision-making process that evaluates and ranks multiple criteria. Unlike traditional single-criterion decision-making methods, MCA can address complex problems that consider multiple dimensions and the perspectives of various stakeholders. This makes it particularly useful in fields such as environmental management, economics, engineering, healthcare, and public policy, where decisions often require balancing different and sometimes conflicting objectives (Belton & Stewart, 2002; Munda, 2004; Keeney & Raiffa, 1976; Figueira et al., 2005).

MCA provides a structured framework for analyzing various decision alternatives based on a set of predefined criteria. This framework enables decision-makers to make more rational, transparent, and justifiable choices by systematically comparing different alternatives. The process typically involves defining the decision framework, identifying relevant criteria, assessing the performance of each alternative against these criteria, and applying a specific MCA method to rank or select the best alternatives.

MCA is essentially an "umbrella" term that includes many different methodologies, each with its strengths and applications. Some indicative methodologies are as follows:

- The Analytical Hierarchy Process (AHP) was developed by Thomas Saaty in the 1980s. It is highly useful when the decision-making process involves multiple criteria that need to be considered, and it is widely used due to its simplicity and its ability to handle both qualitative and quantitative data (Saaty, 1980), as well as to incorporate the subjective judgment of the decision-maker (Subramanian & Ramanathan, 2012). The steps for applying AHP are as follows: The problem

C. Zopounidis, G. Fakotakis, *Multi-Criteria Methodology for the Dynamic Development of Tourist Areas*, Multiple Criteria Decision Making,
https://doi.org/10.1007/978-3-031-95314-9_3

and the goal of the decision-making process are defined. The decision problem is decomposed into hierarchical and more easily understandable sub-problems. The hierarchy typically has at least three levels: the overall decision goal, the criteria used to evaluate the alternatives, and the different available alternative options. Pairwise comparisons of the criteria and alternative solutions are conducted to determine their relative importance or preference. Each element is compared with every other element in terms of its impact on an element above it in the hierarchy. A fundamental scale of absolute numbers (1 to 9) is used to represent how much more one element is preferred over another with respect to a given characteristic (Saaty, 1990). Matrices are then constructed for each level of the hierarchy, where each entry in the matrices represents the result of a pairwise comparison. These matrices are used to derive priority scales, which are normalized to reflect the relative priorities of the criteria and alternatives. The consistency ratio (CR) is used to evaluate how consistent the results are and ensures that the pairwise comparisons are relatively consistent. Finally, the relative priorities of the criteria and alternative solutions are combined to determine an overall priority ranking. This involves aggregating the weights assigned to each criterion and alternative to arrive at an overall ranking of the alternatives. AHP offers many advantages, including its structured and systematic approach to decision-making, which facilitates the handling of complex problems with multiple criteria (Tavana et al., 2023).

- Multi-Attribute Utility Theory (MAUT) evaluates alternatives by focusing on utility functions that represent the preferences of decision-makers. It was developed based on utility theory and provides a rigorous framework for making trade-offs between different criteria in decision-making when dealing with complex scenarios involving multiple objectives. To apply MAUT, the decision problem is clearly defined, and the goals and criteria for evaluating alternatives are identified. A utility function is assigned to each criterion, which quantifies the decision-maker's preferences for different performance levels related to that criterion (Ananda & Herath, 2005). To create the utility functions, preferences are either directly obtained from decision-makers through questionnaires or interviews, or historical data is analyzed. Once the utility functions are established, each criterion is weighted based on its relative importance in the decision-making process. The weights can be derived through various methods, such as pairwise comparisons or direct assessment techniques. The weighted utility values for each criterion are aggregated to calculate an overall utility score for each alternative. MAUT can handle both quantitative and qualitative data, making it suitable for a wide range of decision-making contexts. The utility functions can be tailored to incorporate risk preferences and other subjective preferences of the decision-maker, enabling precise and customized evaluations of alternative options (Keeney & Raiffa, 1976).
- TOPSIS (Technique for Order of Preference by Similarity to Ideal Solution) is based on comparing each alternative to an ideal solution. The core principle for its application is that the selected alternative should have the shortest geometric distance from the best possible solution and the greatest geometric distance from

the worst possible solution (Hwang & Yoon, 1981). To apply TOPSIS, the decision matrix is first constructed, which includes all alternative solutions and criteria. Each element of the matrix represents the performance of an alternative with respect to a criterion. The decision matrix is normalized so that the different scales of criteria can be comparable. After normalization, weights are assigned to each criterion based on their relative importance, and the normalized decision matrix is multiplied by these weights to create the weighted normalized decision matrix. Next, the best and worst possible solutions are identified. The best possible solution represents the optimal value for each criterion, while the worst possible solution represents the poorest value for each criterion. These "extreme" solutions serve as reference points against which all alternatives are compared. Subsequently, separation measures are calculated based on the Euclidean distance of each alternative from the best and worst possible solutions, indicating how close each alternative is to these points. For each alternative, the relative closeness to the best solution is calculated by dividing the separation measure from the worst solution by the sum of the separation measures from both the best and worst solutions. Finally, the alternatives are ranked based on their relative closeness to the best solution, with the alternative having the highest relative closeness being considered the best choice. TOPSIS offers several advantages, including a clear and simple computational process, making it easy to understand and apply. The method considers both the best and worst possible scenarios, providing a balanced evaluation of alternatives. Additionally, it can handle both qualitative and quantitative data, making it useful in a wide range of decision-making contexts (Pandey et al., 2023).
- The ELECTRE (Elimination and Choice Expressing Reality) method uses dominance relations to compare alternatives based on multiple criteria (Roy, 1968). Unlike other multi-criteria analysis methods that seek to find an overall optimal solution, ELECTRE focuses on identifying and comparing alternatives that are "worse" or preferable to others according to a number of criteria (Roy, 1968). To apply the method, initially, the criteria and alternatives are defined. It is important to identify a comprehensive and relevant set of criteria and alternatives for the problem we are faced with. Then, a decision matrix is constructed where each element represents the performance of an alternative with respect to a criterion. The decision matrix is normalized so that the criteria can be compared, using various normalization techniques such as linear normalization. After normalization, weights are assigned to the criteria according to their relative importance, reflecting the priority given to the particular criterion or the importance of each criterion in the decision-making process. For each pair of alternatives, the agreement and disagreement indices are calculated. The agreement index measures the degree to which one alternative is at least as good as another, while the disagreement index measures the degree to which one alternative is worse than another based on any criterion. Using these indices, superiority relations between pairs of alternatives are determined. One alternative is superior to the other if the agreement index is high and the disagreement index is low. A superiority matrix is then constructed to summarize the superiority relations for all pairs of alterna-

tives. Finally, the best alternatives are identified, which can be done by constructing a graph to visualize the higher-ranking relations and facilitate the selection process. ELECTRE is particularly effective in handling complex decision-making problems with conflicting criteria. The method is flexible and has the ability to evaluate qualitative and quantitative data, as well as imprecise and uncertain information, which makes it a powerful tool in the real world. Variants of the method (ELECTRE II, III, etc.) are used when additional procedures (such as ranking) are required to address a problem (Roy & Vincke, 1981).

- PROMETHEE (Preference Ranking Organization Method for Enrichment Evaluations) is similar to the ELECTRE method and is based on precedence relations, but uses preference functions to rank alternatives (Brans & Vincke, 1985). To apply the method, the criteria and alternatives must be defined. For each criterion, a preference function is defined based on the nature of the criterion and the preferences of the decision maker. A performance matrix is constructed, where each element represents the performance of an alternative with respect to a criterion. This matrix is used to perform pairwise comparisons of alternatives for each criterion. The corresponding preference functions are applied to calculate preference indices, which reflect the degree to which one alternative is preferred over another. After calculating the preference indices, the cumulative preference index is calculated for each pair of alternatives and is used to construct preference flows for each alternative. As is the case with the ELECTRE method, there are variants of PROMETHEE that are used depending on the problem being addressed. For example, PROMETHEE I gives us a partial ranking and can be used to identify clear preferences as well as alternatives that are not clearly superior to each other, while PROMETHEE II provides an overall ranking (Brans & Vincke, 1985). PROMETHEE III can be used for interval-based ranking, PROMETHEE IV when dealing with a continuous set of viable solutions for both partial and full ranking, while PROMETHEE V can deal with problems involving segmentation constraints (Taherdoost & Madanchian, 2023)
- The UTA method aims to construct a utility function that best fits a set of reference alternatives provided by the decision maker. The primary goal of the method is to model the decision maker's preferences through an additive utility function. This function can evaluate and rank new alternatives (Jacquet-Lagrèze & Siskos, 1982). Accordingly, there are various variations of the basic model to address various issues that may be required (Beuthe & Scannella, 2001).
- UTADIS is an extension of UTA and focuses on classification rather than ranking, aiming to classify alternatives into predefined groups based on constructed utility functions (Zopounidis & Doumpos, 1999). The user selects the groups according to the problem he faces and the decision he is asked to make. Initially, he should use a reference set which includes alternatives that he will have evaluated and classified into groups according to his preferences and the purpose of the analysis. Then, UTADIS, through linear programming and monotonic regression techniques, develops an additive utility function and calculates the utility limits in order to classify the alternatives into the groups they belong to with the least possible error (Pendaraki et al., 2005). That is, it analyzes the decision-

maker's overall preferences in order to create a model that will synthesize the criteria in the way that the decision-maker makes his decisions. A reference set of alternative solutions is used, which are evaluated based on some criteria in order to classify the alternatives into groups. The UTADIS method classifies the alternatives by giving a total utility score to each of them. This score measures the overall performance of this alternative in all the criteria considered. The marginal utility functions for each criterion take values from 0 for the worst possible alternative, to 1 for the best possible alternative (for criteria that are weighted on an ascending scale). The total score is obtained by summing the marginal utilities weighted by the corresponding weight.

- MACBETH (Measuring Attractiveness by a Categorical Based Evaluation Technique) helps decision makers quantify their preferences for various alternatives without necessarily requiring any numerical input and provides a way to evaluate different options through pairwise comparisons, with the aim of creating a consistent and robust value function (Bana e Costa & Vansnick, 1994). To quantify preferences, a semantic scale is used with categories such as "no preference", "very weak preference", "weak preference", "moderate preference", "strong preference", "very strong preference" and "absolute preference". A consistency check ensures that the qualitative judgments are consistent to verify that the preferences expressed by the decision maker do not contradict each other. The qualitative judgments are then converted into a quantitative value function. This involves solving an optimization problem to derive a set of numerical values that reflect the decision maker's preferences as accurately as possible. A value function is used to evaluate and rank the alternatives. Each alternative is assigned a value based on the decision maker's preferences, allowing for a clear comparison. Sensitivity analysis is also included to understand how changes in the decision maker's preferences or weighting criteria affect the ranking of the alternatives (Ömür, 2017).
- The Simple Multi-Attribute Rating Technique (SMART) is a simple and flexible multi-criteria analysis method that provides a simpler alternative to more complex methods such as AHP and MAUT. SMART uses direct evaluation and linear additive models to facilitate decision making (Rashidi et al., 2018). The decision maker evaluates each alternative directly against each criterion, and then these evaluations are combined using weights that reflect the relative importance of the criteria. (Barfod & Leleur, 2014) This approach is particularly useful due to its simplicity and ease of understanding, making it accessible to a wide range of users without requiring too much knowledge.

In recent years, the use of Multi-Criteria Analysis has expanded due to the flexibility and adaptability of the methods in various decision-making contexts. Multi-Criteria Analysis can, for example, be used, among others:

- In urban and regional planning to optimize land-use decisions and infrastructure development (Thill, 1999) and to create "smart" cities (Aragão et al., 2023).
- In healthcare (Liberatore & Nydick, 2008; Schmidt et al., 2015), Multi-Criteria Analysis can be used to prioritize different medical treatments and policies (Marsh & Dolan, 2014; Sainfort & Gustafson, 1990).

- In energy planning, it can be used to evaluate different energy policies and renewable energy projects (Pohekar & Ramachandran, 2004; Ertay et al., 2013; Montignac et al., 2009; Wu et al., 2018; Mokarram et al., 2020).
- In supply chain analysis (Chan & Kumar, 2007). Topics addressed here include improving competitiveness, optimizing business processes, maximizing revenue and profits, better inventory management, improving customer satisfaction, and reducing costs (Chen & Hung, 2012). It can also be used to evaluate sustainable management issues in companies operating in this sector (Govindan et al., 2013) and in the selection of "green" suppliers (Abdullah et al., 2019).
- In business management for the evaluation of companies (Deng et al., 2000) and corporate sustainability (Bezerra et al., 2021).
- In human resource management (Shih et al., 2007).
- In transportation (Özceylan et al., 2016; Tadić et al., 2019; Fartaj et al., 2019; Rezaei et al., 2017; Tsita & Pilavachi, 2012).
- In the stock market (Albadvi et al., 2007).
- In finance (Dimitras, 2002; Ghanbarizadeh et al., 2019).
- In country risk assessment (Kosmidou et al., 2008).
- In public works assessment (Bana e Costa & Oliveira, 2002).
- In water quality assessment and management (Singh & Singh, 2022; Sutadian et al., 2017; Kilinç et al., 2018; Mandal & Mandal, 2018; Seejata et al., 2018).
- In natural disaster prevention and response (Romario et al., 2018, Kayastha et al., 2013,).
- In the automotive industry (Sapuan et al., 2011, Petruni et al., 2019, Govindan et al., 2013).
- In tourism (Zhang, 2017; Gupta, 2018; Kumar et al., 2020).
- In agricultural production (Giri & Nejadhashemi, 2014; Amini & Asoodar, 2016; Abdollahzadeh et al., 2016; Hafezalkotob et al., 2018).
- In wastewater and waste management (Feyzi et al., 2019; Aung et al., 2019; Ren et al., 2017).
- In construction (Ignatius et al., 2016; Khoshnava et al., 2018)

Multi-criteria analysis is also a useful tool for many environmental problems (Ramanathan, 2001) that present and concern both researchers and local authorities managing these areas. The reason is that decisions related to the environment are generally very complex, having to take into account many different interest groups with different goals and priorities (Munda, 2004). They also have to take into account different criteria such as environmental impact, cost and social acceptance (Hobbs & Meier, 2000). They also often combine problems that require multifaceted scientific knowledge and are called upon to process a large volume of data (Greco et al., 2016) of different formats, incorporating subjective judgments (Bana e Costa & Chagas, 2004) and combination with other methods (Linkov & Moberg, 2012). Depending on the problem, various multi-criteria methods are used, such as the Analytical Hierarchy Process (AHP), PROMETHEE, ELECTRE, etc. (Kiker et al., 2005).

A typical example of the use of multi-criteria analysis in environmental issues is to decide whether an area needs a specific management plan and should be included in the Natura 2000 network of protected areas. A related study published in 2014,

(Cortina & Boggia, 2014) used multi-criteria decision analysis to create biodiversity and human activity indices for various candidate areas. By crossing the 2 indices, it resulted in 4 different categorizations depending on whether they have high or low values in each index, concluding that those with a high biodiversity index and at the same time significant human activity that burdens them are those that need protection most immediately.

A second typical example of the use of multi-criteria methodology is to evaluate the results of various management plans of a protected area according to the expectations of different interest groups operating in these areas in order to make management more effective. A related study published in 2011, (Heck et al., 2011) applied the Analytical Hierarchy Process (AHP) method for the "Southern Strait of Georgia" marine protected area, which converts subjective evaluation into a set of "weights". Data were collected for different possible management actions, coded and local interest groups were asked to make pairwise comparisons between the alternatives in order to reveal their preferences for one alternative over another. The synthesis of the results showed on the one hand, cumulatively, which category of actions is given the greatest importance. However, going deeper, it showed significant differences in preferences both between different actions in the same category and between significantly different preferences depending on the interest group.

The use of multi-criteria methodology is also ideal for measuring and analyzing the satisfaction of a user of a product or service by the company or entity that offers them (customer satisfaction). The advantages offered by such an analysis are very important since it provides the necessary information in order to select the appropriate actions and strategies, so that the provided product or service becomes better and more competitive compared to alternative options (Angilella et al., 2015).

Measuring this satisfaction is considered the most reliable method of evaluating the product since it captures the preferences and expectations of the user-customer with specific qualitative parameters-criteria, which lead to easily understandable results. In order to implement this, it is necessary to evaluate the different dimensions of satisfaction and to separate customer groups with separate preferences. The implementation of the analysis will provide the average overall level of customer satisfaction, the individual levels of satisfaction based on each selected criterion, the specific weights of each criterion, as well as something that will allow a business to identify its strengths and weaknesses (Mihelis et al., 2001).

3.2 The MUSA Method

A popular multi-criteria method for measuring customer satisfaction is the Multicriteria Satisfaction Analysis (MUSA). It was developed at the Laboratory of Decision Support Systems Design and Development at the Technical University of Crete. Its main goal is to combine the preferences of a group of customers into a quantitative mathematical value function. The fundamental assumption of the method is that the overall satisfaction of a customer depends on their satisfaction with specific characteristics (variables-criteria) of the service or product being evaluated, such that X= (X1, X2, …, Xn) as presented in Fig. 3.1.

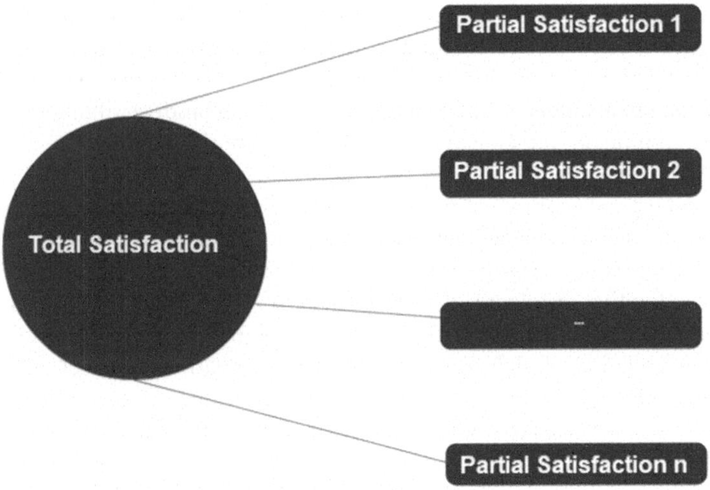

Fig. 3.1 From Partial to Overall Satisfaction

The method uses data collected through properly designed questionnaires completed by customers, where they evaluate various parameters of a business, service, or product. It relies on categorical regression techniques. In such a questionnaire, each customer is asked to evaluate a "product" overall and to assess individual characteristics (criteria) that are considered to contribute to overall satisfaction with the product.

MUSA considers the values and preferences stated by the sample of customers under study and evaluates the satisfaction level of a group of customers by aggregating the various preferences into distinct satisfaction functions, aiming to minimize possible deviations (Grigoroudis & Siskos, 2010).

3.2.1 Assumptions of the MUSA Method

To apply the MUSA method, the following assumptions are made:

1. Rational Consumer Behavior: Consumers make decisions based on logical reasoning.
2. Satisfaction Criteria: Customer satisfaction is based on a predefined set of characteristics.
3. Additive Aggregation Model: This relates to an additive value function that adheres to the property of monotonicity within a multi-criteria decision-making framework.

The MUSA approach assumes both an overall satisfaction function, denoted as Y^*, and a series of partial satisfaction functions, represented as Xi^*. The variables that are used are presented in Table 3.1.

Table 3.1 Variables of the MUSA method

Variable	Description
Y	Total satisfaction
α	Number of levels on the overall satisfaction scale
y^m	m-th level of overall satisfaction (m = 1,2,…, α)
n	Number of criteria
X_i	Customer satisfaction with criterion i (i = 1,2,…,n)
$α_i$	Number of levels on the satisfaction scale for criterion i
x_i^k	k-th satisfaction level for criterion i (k = 1,2,…,$α_i$)
Y^*	Total satisfaction function
y^{*m}	Value of y^m-th satisfaction level
X_i^*	Partial satisfaction function
x_i^{*k}	Value of xi^k-th satisfaction level

MUSA operates similarly to constrained regression analysis while utilizing linear programming techniques. Therefore, the following equation applies:

$$Y^* = \sum_{i=1}^{n} b_i X_i^* \; with \sum_{i=1}^{n} b_i = 1$$

The weights have a total sum of one, while the overall and partial satisfaction functions are monotonic and normalized within the range [0, 100].

Thus, for Y* and Xi* the following holds:

$$\left\{ \begin{array}{c} y^{*1} = 0, y^{*a} = 100 \\ x^{*1} = 0, x^{*a} = 100 \, for \; i = 1,2,\ldots,n \end{array} \right\}$$

The preference relationships model the monotonicity constraints of the overall and partial satisfaction functions:

$$\left\{ \begin{array}{c} y^{*m} \le y^{*m+1} \Leftrightarrow \pm y^m \le y^{m+1} \; for \; m = 1,2,\ldots,n \\ x_i^{*k} \le x_i^{*k+1} \Leftrightarrow x_i^k \le x_i^{k+1} \; for \; k = 1,2,\ldots,a_i - 1 \kappa \alpha \iota \, i = 1,2,\ldots,n \end{array} \right.$$

The primary goal of MUSA is to maximize consistency between these functions. The satisfaction functions, Xi*, represent customer satisfaction regarding a specific characteristic of the product or service, while Y* represents the overall satisfaction of a group of customers. The method incorporates a dual-error variable into the categorical regression calculation:

$$\tilde{Y}^* = \sum_{i=1}^{n} b_i X_i^* - \sigma^+ + \sigma^-$$

In this equation:

$\tilde{Y}^*$ is the estimation of the total satisfaction function Y^*
n is the number of the satisfaction criteria that we use in the model
b_i is the weight of criteria i and also $\sum_{i=1}^{n} b_i = 1$
σ^+ is the overestimation error
σ^- is the underestimation error.

The estimation of customer satisfaction can be formulated as a mathematical programming problem, aiming to minimize the sum of error variables under the following constraints:

1. Additive Utility Function: Given a set of criteria ($g_1, g_2, \ldots, g_n$) with n > 3, there is an additive utility function if mutual independence of preferences holds for the set of criteria:

$$u(g_1, g_2, \ldots, g_n) = \sum_{i=1}^{u} u_i(g_i)$$

2. Normalization Constraints: The Y^* and Xi^* functions are normalized within the range [0, 100].
3. Monotonicity Constraints: The Y^* and Xi^* functions must adhere to monotonicity.

3.2.1.1 Simplifying the Optimization Problem

To simplify the problem and reduce computational difficulty in finding the optimal solution, monotonicity constraints can be eliminated. This is achieved by introducing new variables to express successive incremental steps in the overall and partial satisfaction functions. The variables are defined as follows:

$$\begin{cases} Z_m = y^{*m+1} - y^{*m} \; for \; m = 1,2,\ldots,a-1 \\ w_{ik} = b_i x_i^{*k+1} - b_i x_i^{*k} \; for \; k = 1,2,\ldots,a_t - 1 \; \kappa\alpha\iota \; i = 1,2,\ldots,n \end{cases}$$

Based on these equations, the initial decision variables of the linear program become:

$$\begin{cases} y^{*m} = \sum_{t=1}^{m-1} Z_t \; for \; m = 2,3,\ldots\alpha \\ b_i x_i^{*k} = \sum_{t=1}^{k-1} w_{it} \; for \; k = 2,3,\ldots,a_t - 1 \; and \; i = 1,2,\ldots,n \end{cases}$$

Using the above equations and the new variables Z_m and w_{ik}, the regression equation becomes:

$$\sum_m Z_m = \sum_i \sum_k w_{ik} - \sigma^+ + \sigma^-$$

3.2.1.2 Customer Satisfaction Expression

If the overall satisfaction of a customer j is expressed based on predefined qualitative scales Y and X such that:

$$\begin{cases} Total\ satisfaction\ \overline{y}^j = y^{t_j}\ and\ \overline{y}^j \in \left\{ y^1, y^2, \ldots, y^{t_j}, \ldots, y^a \right\} \\ Partial\ satisfaction\ \overline{x}_i^j = x_i^{t_\mu}\ and\ \overline{x}_i^j \in \left\{ x_i^1, x_i^2, \ldots, x_i^{t_\mu}, \ldots, x_i^a \right\}\ for\ i = 1,2,\ldots,n \end{cases}$$

Then, the following holds:

$$\sum_m^{t_j - 1} Z_m = \sum_{i=1}^{m} \sum_{k=1}^{t_{ij}-1} w_{ik} - \sigma^+ + \sigma^-\ for\ every\ j$$

The linear program takes the following final form:

$$\begin{cases} [\min] F = \sum_{j=1}^{M} \sigma_j^+ + \sigma_j^-\ subject\ to: \\ \sum_{i=1}^{n} \sum_{k=1}^{t_j - 1} Z_m - \sigma_j^+ + \sigma_j^- = 0\ \gamma\iota\alpha\ j = 1,2,\ldots,M \\ \sum_{m=1}^{a-1} Z_m = 100 \\ \sum_{i=1}^{n} \sum_{k=1}^{a_i - 1} w_{ik} = 100 \\ Z_m \geq 0, w_{ik} \geq 0\ for\ every\ m,i,k \\ \sigma_j^+ \geq 0, \sigma_j^- \geq 0\ for\ j = 1,2,\ldots,M \end{cases}$$

Where M is the total number of customers.

The reliability of the results is evaluated using the adjustment and stability index (Grigoroudis & Siskos, 2010). Among the advantages of MUSA is the fact that it provides a clear picture of customer satisfaction, allowing us to understand why the customer is satisfied or dissatisfied, and also to identify the strengths and weaknesses of the product. MUSA can support customer-related decisions by recognizing, for example, the product characteristics that need priority attention and the optimal way to improve satisfaction. It also enables us to develop an integrated management system that emphasizes customer preferences to enhance the product.

The information obtained after applying MUSA determines the dimensions on which the business should focus to improve (Grigoroudis & Siskos, 2003). This information provides us with the overall level of customer satisfaction (general consumer satisfaction), the individual satisfaction levels for each criterion, and the weights for each criterion (criteria weights and satisfaction indices) that contribute to the overall level. It also allows for the construction of action diagrams and improvement diagrams.

Specifically, MUSA provides us with:

1. General Consumer Satisfaction Level: This indicates the overall level of satisfaction on a 0–100% scale, showing how satisfied the customer is with the product overall.
2. Satisfaction Functions: These are critical as they demonstrate the actual value customers assign to a defined qualitative satisfaction level. Based on these functions, the level of customer demand can be determined. Customers are categorized into three main groups:

   - Non-demanding Customers: Satisfied even if not all expectations are met. The satisfaction function is concave.
   - Neutral Customers: Satisfaction increases linearly with the percentage of expectations fulfilled. The satisfaction function is linear.
   - Demanding Customers: Satisfied only when all expectations are met. The satisfaction function is convex.

3. Criterion Satisfaction Indices: These indicate the average satisfaction of customers with a specific product feature and are expressed on a 0–100% scale.

   The mean satisfaction index S and partial satisfaction indices S_i are calculated based on the following relations:

$$\begin{cases} S = \dfrac{1}{100}\sum_{m=1}^{\alpha} p^m y^{*m} \\ S_i = \dfrac{1}{100}\sum_{k=1}^{a_i} p_i^k x_i^{*k} \; for\ i = 1,2,\ldots,n \end{cases}$$

   Where p^m and p_i^k represent the percentage of customers belonging to the satisfaction levels y^m and x_i^k, respectively. The mean indices are normalized within the interval [0–100] and represent the average value of the corresponding satisfaction function for the specific criterion.

4. The criteria weights indicate the relative importance of a specific criterion based on the preferences of the customer group and can also be interpreted as trade-offs between their respective values.

5. The average demandingness indices are a quantitative measure for the concept of demandingness. They take into account all information derived from the satisfaction functions and indicate how demanding the customer is concerning a specific feature (criterion).

   The overall (D) and partial (D_i) demandingness indices are calculated according to the following relationships:

$$\begin{cases} D = \dfrac{\sum_{m=1}^{a-1}\left(\dfrac{100(m-1)}{a-1} - y^{*m}\right)}{100\sum_{m=1}^{a-1}\dfrac{m-1}{a=1}} \quad for\,\alpha > 2 \\ D_i = \dfrac{\sum_{k=1}^{\alpha_i-1}\left(\dfrac{100(k-1)}{a_i-1} - x_i^{*k}\right)}{100\sum_{k=1}^{a_i-1}\dfrac{k-1}{a_i-1}} \quad for\,\alpha_i > 2\,and\,i = 1,2,\ldots,n \end{cases}$$

   These indices are normalized, expressed on a scale from −1 to 1, and essentially represent the average deviation of the satisfaction functions from a linear satisfaction function. They help to understand the level of effort required to improve a specific feature, as the more demanding the customer is, the greater the effort needed to meet their expectations.

   The following applies:

- For D = 1 or D_i = 1, the maximum degree of demandingness is observed.
- For D = 0 or D_i = 0, customers are considered "neutral."
- For D = −1 or D_i = −1, the minimum degree of demandingness is observed.

Since the indices rarely take extreme values, and to enable meaningful conclusions, the scale from −1 to 1 can be divided into three categories:

- Non-demanding (−1 to −0.33): In this category, the indices show that significant effort for improvement is not required. Criteria in this range may already be in good condition or do not require substantial changes.
- Neutral (−0.33 to 0.33): These indices indicate a moderate level of effort required.
- Demanding (0.33 to 1): In this category, the indices suggest that significant effort is required for improvement. These areas may need substantial changes or enhancements to reach satisfactory levels.

6. The average effectiveness indices show how effective a potential improvement in a specific feature will be. This effectiveness depends on the importance of the specific feature for customers and the potential for improvement. These indices can prioritize improvement actions and are calculated as follows:

$$I_i = b_i\left(1 - S_i\right) for\, i = 1,2,\ldots,n$$

7. The action diagram (Fig. 3.2) is created based on the performance of a criterion, shown on the vertical axis, and the evaluation of its importance, displayed on the horizontal axis.

   By using the absolute weights (b_i) and partial satisfaction indices (S_i), an absolute action diagram is constructed. The importance axis takes values within the interval [0, 1]. A criterion is considered significant if $b_i > 1/n$, as the weight of each criterion depends on the number n of criteria. The performance axis takes values within the interval [0, 1], based on the average partial satisfaction indices (S_i). Customers are considered satisfied with a criterion if it receives a score greater than 0.5 (50%), though this assumption is subjective and can be adjusted if necessary.

   The relative action diagram is used to resolve issues such as determining the intersection points with the horizontal/vertical axis in the absolute action diagram or addressing small variations in the average satisfaction indices. For its creation, normalized variables for weight (b_i') and satisfaction (S_i') are used and are calculated as follows:

$$b_i' = \frac{b_i - \overline{b}}{\sqrt{\sum_i (b_i - \overline{b})^2}}$$

$$S_i' = \frac{S_i - \overline{S}}{\sqrt{\sum_i (S_i - \overline{S})^2}}$$

Where $\overline{b}$ and $\overline{S}$ are the mean values of weights and satisfaction indices, respectively.
The values of b_i' and S_i' ′ fall within the interval [−1, 1].

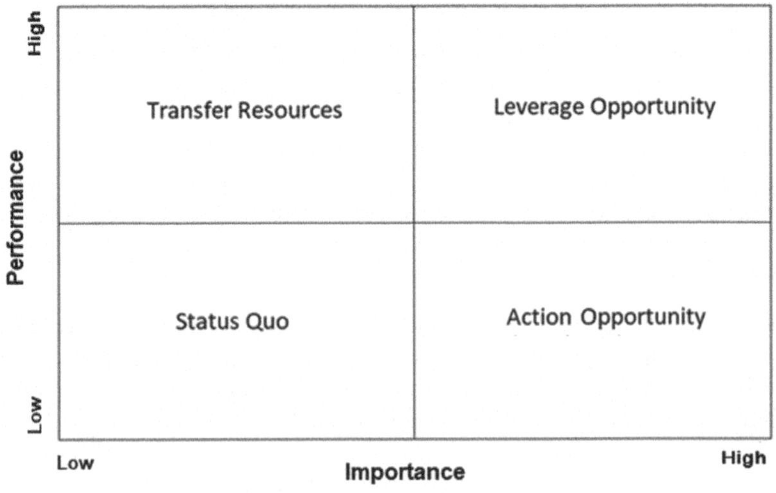

Fig. 3.2 Action diagram

The action diagram is divided into four quadrants: status quo, leverage opportunity, transfer resources, and action opportunity (Grigoroudis & Siskos, 2010).

- Status quo: Combines low performance and low importance. No action is required, as satisfaction is not deemed critical.
- Leverage opportunity: Characterized by high performance and high importance. These represent competitive advantages, as their satisfaction significantly influences purchasing decisions.
- Transfer resources: Combines high performance and low importance. Resources allocated here could be redirected to areas of greater importance to customers.
- Action opportunity: Combines low performance and high importance, requiring focused attention to improve performance and enhance customer satisfaction.

It is obvious that an effort should be made to improve the characteristics in the action opportunity area first, since these are criteria that are considered important but with which customers are less satisfied. Next, attention should be given to the characteristics in the Leverage Opportunity area when there is room for improvement. Third in priority are the characteristics in the Status Quo area. These characteristics are not particularly important but may become important in the future and it should be understood that customer satisfaction with them is low. Last priority is given to the characteristics in the Transfer Resources area since they do not seem to be important to customers while their performance is quite high.

8. The improvement diagram (Fig. 3.3) is constructed to address the problem arising from the action diagram, which can indicate which criteria need to be improved but cannot show what the result of the improvement will be, and the amount of effort needed to achieve it.

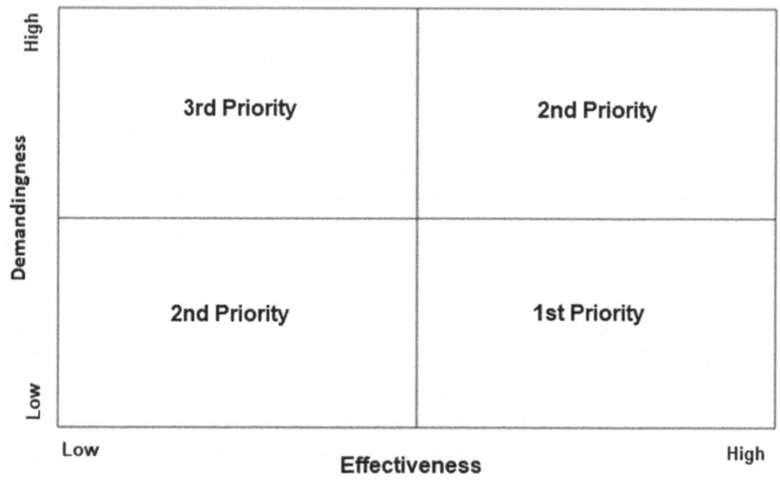

Fig. 3.3 Improvement diagram

It combines the average demand indicators and the average effectiveness indicators. It consists of assessing how demanding an action is on the vertical axis and assessing the impact that this action has on the horizontal axis. It is divided into quadrants according to the level of demandingness and the level of effectiveness in order to determine the priorities for improvement.

First priority is given to improvements that will have high efficiency and low demandingness. Second in priority are improvement actions that are either characterized by high efficiency and high demandingness, or by low efficiency and low demandingness. Last come the characteristics that are low efficiency and high demandingness. Therefore, depending on the point at which each criterion is located in a diagram, the MUSA method reveals the strengths and weaknesses of the company under analysis as well as the opportunities for future action (Manolitzas et al., 2014). Similar to the action diagram, an absolute or relative action diagram can be constructed. In the absolute action diagram, the Efficiency indicators Ii are used on the horizontal axis, which takes values in the interval [0, 1], while the demandingness indicators Di are used on the vertical axis, which takes values in the interval [−1, 1].

To create the relevant improvement diagram, normalized variables for efficiency (I_i') and for demandingness $\left(D_i'\right)$ are used and are calculated as follows:

$$I_i' = \frac{I_i - \overline{I}}{\sqrt{\sum_i (I_i - \overline{I})^2}}$$

$$D_i' = \frac{D_i - \overline{D}}{\sqrt{\sum_i (D_i - \overline{D})^2}}$$

Where $\overline{I}$ and $\overline{D}$ are the average values of the efficiency indices and the demand indices, respectively.

I_i' and D_i' take values in the interval [−1, 1]

9. The fit and stability indices check the reliability of the results and the interpretability of the MUSA method on the satisfaction survey data. The average fit index AFI_1 is a normalized index and is calculated as follows:

$$AFI_1 = 1 - \frac{F^*}{100M}$$

Where M is the number of customers and F^* is the optimal error value of the original linear program. It takes values in the interval [0, 1]. Specifically, the index takes the value 1 when the model is able to estimate the total values for customers, while on the contrary it takes the value 0 when the error variables σ_i^+ and σ_i^- take their maximum possible value. The average fit index AFI2 calculates the percentage of customers with zero error. That is:

$$AFI_2 = \%\ \text{of customers with}\ \sigma^+ = \sigma^- = 0$$

The average fit index AFI_3 is an improved version of the AFI_1 index and distinguishes each level of satisfaction of the overall satisfaction and calculates the maximum error that can exist for this level.

It is calculated as:

$$AFI_3 = 1 - \frac{F^*}{M \sum_{m=1}^{a} P^m \max\left\{y^{*m}, 100 - y^{*m}\right\}}$$

The average stability index ASI is the average value of the normalized standard deviation of the estimated weights bi for the criteria of the problem. It is calculated as follows:

$$ASI = 1 - \frac{1}{n} \sum_{i=1}^{n} \frac{\sqrt{n \sum_{j=1}^{n} \left(\sum_{j=1}^{n} b_i^j \right)^2}}{100\sqrt{n-1}}$$

Where b_i^j is the estimated weight of criterion i when solving the j linear program during meta-optimization. ASI takes values between 0 and 1.

The MUSA method can be applied to many different business cases. Typical examples from the literature are the banking sector (Mihelis et al., 2001), the evaluation of citizens' satisfaction with KEPs (Manolitzas & Yannacopoulos, 2013), the medical sector (Manolitzas et al., 2014) and in any case of customer satisfaction evaluation in general (Siskos & Grigoroudis, 2002). Therefore, it is ideal to use it for the evaluation of visitor satisfaction at the Samaria National Park.

References

Abdollahzadeh, G., Damalas, C. A., Sharifzadeh, M. S., & Ahmadi-Gorgi, H. (2016). Selecting strategies for rice stem borer management using the Analytic Hierarchy Process (AHP). *Crop Protection, 84*, 27–36.

Abdullah, L., Chan, W., & Afshari, A. (2019). Application of PROMETHEE method for green supplier selection: A comparative result based on preference functions. *Journal of Industrial Engineering International, 15*(1), 271–285.

Albadvi, A., Chaharsooghi, S. K., & Esfahanipour, A. (2007). Decision making in stock trading: An application of PROMETHEE. *European Journal of Operational Research, 177*(2), 673–683.

Amini, S., & Asoodar, M. A. (2016). Selecting the most appropriate tractor using Analytic Hierarchy Process—An Iranian case study. *Information Processing in Agriculture, 3*(4), 223–234.

Ananda, J., & Herath, G. (2005). Evaluating public risk preferences in forest land-use choices using multi-attribute utility theory. *Ecological Economics, 55*(3), 408–419.

Angilella, S., Corrente, S., Greco, S., & Slowinski, R. (2015). MUSA-INT: Multicriteria customer satisfaction analysis with interacting criteria. *Omega, 42*(1), 189–200.

Aragão, F. V., Chiroli, D. M. d. G., Zola, F. C., Aragão, E. V., Marinho, L. H. N., Correa, A. L. C., & Colmenero, J. C. (2023). Smart cities maturity model—A multicriteria approach. *Sustainability, 15*(8), 6695.

Aung, T. S., Luan, S., & Xu, Q. (2019). Application of multi-criteria-decision approach for the analysis of medical waste management systems in Myanmar. *Journal of Cleaner Production, 222*, 733–745.

Bana e Costa, C. A., & Chagas, M. P. (2004). A career choice problem: An example of how to use MACBETH to build a quantitative value model based on qualitative value judgments. *European Journal of Operational Research, 153*, 323.

Bana e Costa, C. A., & Oliveira, C. S. (2002). Assigning priorities for maintenance, repair and refurbishment in managing a municipal housing stock. *European Journal of Operational Research, 138*(2), 380–391.

Bana e Costa, C. A., & Vansnick, J. C. (1994). MACBETH—An interactive path towards the construction of cardinal value functions. *International Transactions in Operational Research, 1*(4), 489–500.

Barfod, M. B., & Leleur, S. (Eds.). (2014). *Multi-criteria decision analysis for use in transport decision making* (2nd ed.). DTU Transport.

Belton, V., & Stewart, T. J. (2002). *Multiple criteria decision analysis: An integrated approach*. Kluwer Academic Publishers.

Beuthe, M., & Scannella, G. (2001). Comparative analysis of UTA multicriteria methods. *European Journal of Operational Research, 130*(2), 246–262.

Bezerra, P. R. S., Schramm, F., & Schramm, V. B. (2021). A multicriteria model, based on the PROMETHEE II, for assessing corporate sustainability. *Clean Technologies and Environmental Policy, 23*, 2927–2940.

Brans, J. P., & Vincke, P. (1985). A preference ranking organization method (the PROMETHEE method for multiple criteria decision-making). *Management Science, 31*, 647.

Chan, F. T. S., & Kumar, N. (2007). Global supplier development considering risk factors using fuzzy extended AHP-based approach. *Omega, 35*(4), 417–431.

Chen, T. Y., & Hung, C. C. (2012). A novel multiple criteria decision-making model for building material supplier selection based on entropy-AHP weighted TOPSIS. *International Journal of Architecture, Engineering and Construction, 1*(3), 192–201.

Cortina, C., & Boggia, A. (2014). Development of policies for Natura 2000 sites: A multi-criteria approach to support decision makers. *Journal of Environmental Management, 141*, 138–145.

Deng, H., Yeh, C. H., & Willis, R. J. (2000). Inter-company comparison using modified TOPSIS with objective weights. *Computers & Operations Research, 27*(10), 963–973.

Dimitras, A. I. (2002). Evaluation of Greek construction companies' securities using UTADIS method. *European Research Studies, V*(1–2), 95–107.

Ertay, T., Kahraman, C., & Kaya, İ. (2013). Evaluation of renewable energy alternatives using MACBETH and fuzzy AHP multicriteria methods: The case of Turkey. *Technological and Economic Development of Economy, 19*(1), 38–62.

Fartaj, S. R., Kabir, G., Eghujovbo, V., Ali, S. M., & Paul, S. K. (2019). Modeling transportation disruptions in the supply chain of automotive parts manufacturing company. *International Journal of Production Economics, 222*, 107511.

Feyzi, S., Khanmohammadi, M., Abedinzadeh, N., & Aalipour, M. (2019). Multi-criteria decision analysis FANP based on GIS for siting municipal solid waste incineration power plant in the north of Iran. *Sustainable Cities and Society, 47*, 101513.

Figueira, J., Greco, S., & Ehrgott, M. (2005). *Multiple criteria decision analysis: State of the art surveys*. Springer.

Ghanbarizadeh, A., Heydari, J., Razmi, J., & Bozorgi-Amiri, A. (2019). A purchasing portfolio model for the commercial construction industry: A case study in a mega mall. *Production Planning & Control, 30*(15), 1283–1304.

Giri, S., & Nejadhashemi, A. P. (2014). Application of analytical hierarchy process for effective selection of agricultural best management practices. *Journal of Environmental Management, 132*, 165–177.

Govindan, K., Khodaverdi, R., & Jafarian, A. (2013). A fuzzy multi criteria approach for measuring sustainability performance of a supplier based on triple bottom line approach. *Journal of Cleaner Production, 47*, 345–354.

Greco, S., Ehrgott, M., & Figueira, J. R. (2016). *Multiple criteria decision analysis: State of the art surveys*. Springer.

Grigoroudis, E., & Siskos, Y. (2003). MUSA: A decision support system for evaluating and analysing customer satisfaction. In K. Margaritis & I. Pitas (Eds.), *Proceedings of the 9th Panhellenic conference in informatics* (pp. 113–127). TEI of Thessaloniki.

Grigoroudis, E., & Siskos, Y. (2010). *Customer satisfaction evaluation: Methods of measuring and implementing service quality*. Springer.

Gupta, H. (2018). Evaluating service quality of airline industry using hybrid best worst method and VIKOR. *Journal of Air Transport Management, 68*, 35–47.

Hafezalkotob, A., Hami-Dindar, A., Rabie, N., & Hafezalkotob, A. (2018). A decision support system for agricultural machines and equipment selection: A case study on olive harvester machines. *Computers and Electronics in Agriculture, 148*, 207–216.

Heck, N., Dearden, P., & McDonald, A. (2011). A stakeholders' expectations towards a proposed marine protected area: A multi-criteria analysis of MPA performance criteria. *Ocean & Coastal Management, 54*, 687–695.

Hobbs, B. F., & Meier, P. (2000). *Energy decisions and the environment: A guide to the use of multicriteria methods*. Springer.

Hwang, C. L., & Yoon, K. (1981). *Multiple attribute decision making: Methods and applications*. Springer.

Ignatius, J., Rahman, A., Yazdani, M., Šaparauskas, J., & Haron, S. H. (2016). An integrated fuzzy ANP–QFD approach for green building assessment. *Journal of Civil Engineering and Management, 22*(4), 551–563.

Jacquet-Lagrèze, E., & Siskos, Y. (1982). Assessing a set of additive utility functions for multicriteria decision-making, the UTA method. *European Journal of Operational Research, 10*(2), 151–164.

Kayastha, P., Dhital, M. R., & De Smedt, F. (2013). Application of the analytical hierarchy process (AHP) for landslide susceptibility mapping: A case study from the Tinau watershed, West Nepal. *Computers & Geosciences, 52*, 398–408.

Keeney, R. L., & Raiffa, H. (1976). *Decisions with multiple objectives: Preferences and value tradeoffs*. Wiley.

Khoshnava, S. M., Rostami, R., Valipour, A., Ismail, M., & Rahmat, A. R. (2018). Rank of green building material criteria based on the three pillars of sustainability using the hybrid multi criteria decision making method. *Journal of Cleaner Production, 173*, 82–99.

Kiker, G. A., Bridges, T. S., Varghese, A., Seager, T. P., & Linkovjj, I. (2005). Application of multicriteria decision analysis in environmental decision making. *Integrated Environmental Assessment and Management, 1*(2), 95–108.

Kilinç, Y., Özdemir, Ö., Orhan, C., & Firat, M. (2018). Evaluation of technical performance of pipes in water distribution systems by Analytic Hierarchy Process. *Sustainable Cities and Society, 42*, 13–21.

Kosmidou, K., Doumpos, M., & Zopounidis, C. (2008). *Country risk evaluation methods and applications*. Springer Science + Business Media, LLC.

Kumar, A., Aswin, A., & Gupta, H. (2020). Evaluating green performance of the airports using hybrid BWM and VIKOR methodology. *Tourism Management, 76*, 103941.

Liberatore, M. J., & Nydick, R. L. (2008). The Analytic Hierarchy Process in medical and health care decision making: A literature review. *European Journal of Operational Research, 189*(1), 194–207.

Linkov, I., & Moberg, E. (2012). *Multi-criteria decision analysis: Environmental applications and case studies*. CRC Press.
Mandal, B., & Mandal, S. (2018). Analytical hierarchy process (AHP) based landslide susceptibility mapping of Lish river basin of eastern Darjeeling Himalaya, India. *Advances in Space Research, 62*(11), 3114–3132.
Manolitzas, P., Grigoroudis, E., & Matsatsinis, N. (2014). Using multicriteria decision analysis to evaluate patient satisfaction in a hospital emergency department. *Journal of Health Management, 16*(2), 245–258.
Manolitzas, P., & Yannacopoulos, D. (2013). Citizen satisfaction: A multicriteria satisfaction analysis. *International Journal of Public Administration, 36*, 614–621.
Marsh, K., & Dolan, P. (2014). Multicriteria decision analysis in healthcare decision making. *PharmacoEconomics*.
Mihelis, G., Grigoroudis, E., Siskos, Y., Politis, Y., & Malandrakis, Y. (2001). Customer satisfaction measurement in the private bank sector. *European Journal of Operational Research, 130*(2), 347–360.
Mokarram, M., Mokarram, M. J., Gitizadeh, M., Niknam, T., & Aghaei, J. (2020). A novel optimal placing of solar farms utilizing multi-criteria decision-making (MCDA) and feature selection. *Journal of Cleaner Production, 261*, 121098.
Montignac, F., Noirot, I., & Chaudourne, S. (2009). Multi-criteria evaluation of on-board hydrogen storage technologies using the MACBETH approach. *International Journal of Hydrogen Energy, 34*(10), 4561–4568.
Munda, G. (2004). Social multi-criteria evaluation (SMCE): Methodological foundations and operational consequences. *European Journal of Operational Research, 158*(3), 662–677.
Ömür, T. (2017). Using macbeth method for technology selection in production environment. *American Journal of Data Mining and Knowledge Discovery, 2*(1), 37–41.
Özceylan, E., Erbaş, M., Tolon, M., Kabak, M., & Durğut, T. (2016). Evaluation of freight villages: A GIS-based multi-criteria decision analysis. *Computers in Industry, 76*, 38–52.
Pandey, V., Komal, & Dincer, H. (2023). A review on TOPSIS method and its extensions for different applications with recent development. *Soft Computing, 27*, 18011–18039.
Pendaraki, K., Zopounidis, C., & Doumpos, M. (2005). On the construction of mutual fund portfolios: A multicriteria methodology and an application to the Greek market of equity mutual funds. *European Journal of Operational Research, 163*(2), 462–481.
Petruni, A., Giagloglou, E., Douglas, E., Geng, J., Leva, M. C., & Demichela, M. (2019). Applying Analytic Hierarchy Process (AHP) to choose a human factors technique: Choosing the suitable Human Reliability nalysis technique for the automotive industry. *Safety Science, 119*, 229–239.
Pohekar, S. D., & Ramachandran, M. (2004). Application of multi-criteria decision making to sustainable energy planning—A review. *Renewable and Sustainable Energy Reviews, 8*, 365.
Ramanathan, R. (2001). A note on the use of the Analytic Hierarchy Process for environmental impact assessment. *Journal of Environmental Management, 63*(1), 27–35.
Rashidi, M., Ghodrat, M., Samali, B., & Mohammadi, M. (2018). *Decision support systems*. InTech.
Ren, J., Liang, H., & Chan, F. T. (2017). Urban sewage sludge, sustainability, and transition for Eco-City: Multi-criteria sustainability assessment of technologies based on best-worst method. *Technological Forecasting and Social Change, 116*, 29–39.
Rezaei, J., Hemmes, A., & Tavasszy, L. (2017). Multi-criteria decision-making for complex bundling configurations in surface transportation of air freight. *Journal of Air Transport Management, 61*, 95–105.
Romario, T., Gunawan, V., & Endro, J. (2018). Implementation of the ELECTRE method for determining the location of evacuation of web-based tsunami disaster. *International Journal of Computer Applications, 180*, 25.
Roy, B. (1968). Classement et choix en présence de points de vue multiples (la méthode ELECTRE). *Revue Française d'Informatique et de Recherche Opérationnelle, 2*, 57.
Roy, B., & Vincke, P. (1981). Multicriteria analysis: Survey and new directions. *European Journal of Operational Research, 8*(3), 207–218.

Saaty, T. L. (1980). *The Analytic Hierarchy Process: Planning, priority setting, resource allocation*. McGraw-Hill.

Saaty, T. L. (1990). How to make a decision: The Analytic Hierarchy Process. *European Journal of Operational Research, 48*(1), 9–26.

Sainfort, F., & Gustafson, D. H. (1990). Multiattribute utility models in health care decision making. In L. Breslow (Ed.), *Encyclopedia of public health*. Macmillan.

Sapuan, S. M., Kho, J. Y., Zainudin, E. S., Leman, Z., Ali, B. A., & Hambali, A. (2011). Materials selection for natural fiber reinforced polymer composites using analytical hierarchy process. *Indian Journal of Engineering & Materials Sciences, 18*, 255–267.

Schmidt, K., Aumann, I., Hollander, I., Damm, K., & von der Schulenburg, J. M. G. (2015). Applying the Analytic Hierarchy Process in healthcare research: A systematic literature review and evaluation of reporting. *BMC Medical Informatics and Decision Making, 15*, 112.

Seejata, K., Yodying, A., Wongthadam, T., Mahavik, N., & Tantanee, S. (2018). Assessment of flood hazard areas using analytical hierarchy process over the lower Yom Basin, Sukhothai Province. *Procedia Engineering, 212*, 340–347.

Shih, H. S., Shyur, H. J., & Lee, E. S. (2007). An extension of TOPSIS for group decision making. *Mathematical and Computer Modelling, 45*(7–8), 801–813.

Singh, N., & Singh, K. R. (2022). Application of TOPSIS—A multi criteria decision making approach in surface water quality assessment. In I. Haq, A. S. Kalamdhad, & S. Dash (Eds.), *Environmental degradation: Monitoring, assessment and treatment technologies*. Springer.

Siskos, G., & Grigoroudis, E. (2002). Measuring customer satisfaction for various services using multicriteria analysis. *International Series in Operation Research & Management Science*, 457–482.

Subramanian, N., & Ramanathan, R. (2012). A review of applications of Analytic Hierarchy Process in operations management. *International Journal of Production Economics, 138*(2), 215–241.

Sutadian, A. D., Muttil, N., Yilmaz, A. G., & Perera, B. J. C. (2017). Using the Analytic Hierarchy Process to identify parameter weights for developing a water quality index. *Ecological Indicators, 75*, 220–233.

Tadić, S., Krstić, M., Roso, V., & Brnjac, N. (2019). Planning an intermodal terminal for the sustainable transport networks. *Sustainability, 11*(15), 4102.

Taherdoost, H., & Madanchian, M. (2023). Using PROMETHEE method for multi-criteria decision making: Applications and procedures. *Iris Journal of Economics & Business Management, 1*(1).

Tavana, M., Soltanifar, M., & Santos-Arteaga, F. J. (2023). Analytical hierarchy process: Revolution and evolution. *Annals of Operations Research, 326*(1), 879–907.

Thill, J. C. (1999). *Multicriteria decision analysis in geographic information science*. Springer.

Tsita, K. G., & Pilavachi, P. A. (2012). Evaluation of alternative fuels for the Greek road transport sector using the Analytic Hierarchy Process. *Energy Policy, 48*, 677–686.

Wu, Y., Zhang, B., Xu, C., & Li, L. (2018). Site selection decision framework using fuzzy ANP-VIKOR for large commercial rooftop PV system based on sustainability perspective. *Sustainable Cities and Society, 40*, 454–470.

Zhang, J. (2017). Evaluating regional low-carbon tourism strategies using the fuzzy Delphi-analytic network process approach. *Journal of Cleaner Production, 141*, 409–419.

Zopounidis, C., & Doumpos, M. (1999). A multicriteria decision aid methodology for sorting decision problems: The case of financial distress. *Computational Economics, 14*(3), 197–218.

Chapter 4
The Data

4.1 Number of Visitors

The number of visitors to the Samaria National Park is recorded each year by the OFYPEKA, which is now responsible for the management of the park, and in previous years by the Chania Forestry Directorate. With the exception of the years 2020 and 2021, when the effects of the Covid-19 pandemic were more significant, the park attracts many thousands of visitors during each season, while a maximum of close to 300,000 visitors has been recorded during the years 1989 to 1993.

For the 5 years in which the survey was carried out, from 2018 to 2022, the total attendance is presented in Fig. 4.1.

The question that arises, however, is how satisfied are visitors with their visit and what contributes to it. Are there things that can be improved so that their experience is even better, which will contribute to attracting even more visitors to the park for the benefit of the local economy?

The obvious assumption, judging by the number of visitors and the businesses related to tourism in the surrounding areas of Samaria, is that visitors must probably be very satisfied with the experience they gained. This view is also supported by the fact that a significant percentage of the businesses around the park, especially in Agia Roumeli, operate only during the tourist season when the park is open.

The present study aims not only to present data on the level of satisfaction of visitors to the Samaria National Park but also to determine the factors (criteria) that contribute to this satisfaction. It also aims to identify the profile of the park visitor and the characteristics of their visit and to identify any differences between years and especially before and after the pandemic.

C. Zopounidis, G. Fakotakis, *Multi-Criteria Methodology for the Dynamic Development of Tourist Areas*, Multiple Criteria Decision Making,
https://doi.org/10.1007/978-3-031-95314-9_4

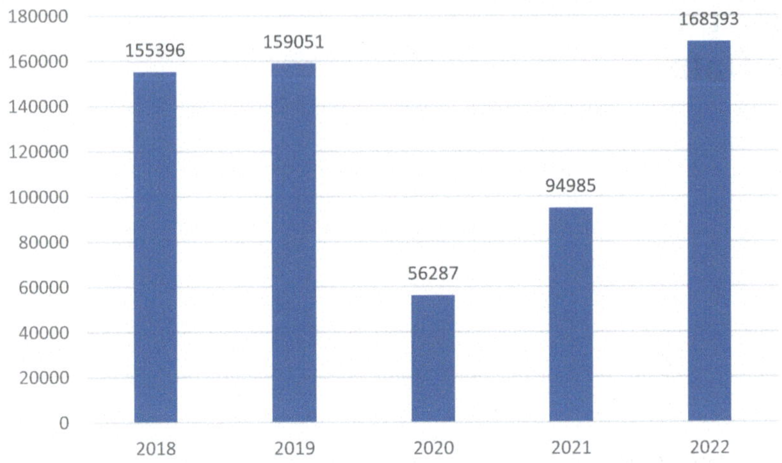

Fig. 4.1 Number of visitors per year

4.2 Data Collection

A sample survey was used to collect the data. An appropriate questionnaire was designed in close collaboration with the then park manager, the Chania Forest Directorate, to be completed by park visitors. Discussions were also held with the staff of the (then) Management Body of the Samaria National Park. The decision for this collaboration was made due to the fact that both the Chania Forest Directorate and the Management Body have direct knowledge of the Samaria National Park. Their knowledge ensured that the questions were relevant and substantive. As those responsible for managing the park, they also have a full understanding of the needs and challenges it faces. Their guidance significantly improved the effectiveness of the research.

This collaboration symbolized a collective commitment to the park and its visitors. It was not just an external survey, but a collective effort with those directly responsible for the management of the park. With their contribution, the questionnaire was better structured to collect information, from visitor preferences to suggestions for improving the park.

The questionnaire was distributed within the Samaria National Park at the rest area of the old village of Samaria. This particular area was chosen because it is the busiest and most comfortable rest area in the park and is located approximately in the middle of the route. At the same time, most visitors choose to stop for a longer period of time at this point. Thus, they have already acquired a good image of the park while at the same time having enough time at their disposal to answer the questionnaire of this study. In collaboration with park officials, visitors were informed about the purposes of the survey and then asked to answer the relevant questions.

The data was collected for the entire duration of the tourist season from May, when the Park opens, to October, when it closes, from 2018 to 2022.

The questionnaire includes three groups of questions. The first group concerns the demographic data of the visitor such as gender, age, level of education and annual income. The aim of these questions was to identify the profile of the visitor to Samaria National Park.

The second group concerns general questions about the visit and the visitors' knowledge about the park. Its aim was to collect data on the information available to visitors about the park, the means of transportation they used and the importance of the park for them.

The third group concerns the visitor's overall satisfaction with their visit and their satisfaction with individual features of the park, which were the criteria for analyzing satisfaction with the MUSA method. The selection of the relevant criteria was not a simple matter.

4.3 The Criteria

There are studies in the literature that show the importance of distinctive features such as the isolation of a destination, local culture and environmental characteristics when it comes to nature tourism (Silverberg et al., 1996). It is common for national parks, as protected areas, to be located in areas with special features, whether they are endemic plants, animals, unique geological formations or specific ecosystems. They offer a peaceful place for visitors to enjoy nature, while preserving biodiversity (Tang & Adesina, 2022). Other studies focus on the role of comprehensive information and effective staff interaction in enhancing the visitor experience. Specifically, the sources of information preferred by visitors are the available bibliography about the site, the existence of educational materials, information points in the site, relevant signage and available information about the biodiversity of the park, while the quality of the park staff is judged by their readiness to help and their kindness (Mcguire et al., 1989). The cleanliness of the areas in which visitors move and the maintenance of infrastructure are also mentioned as factors (Geng et al., 2021) and especially the condition of the trails and rest areas (Zolfaghari & Choi, 2023) while the experiences of visitors within the park, such as wildlife sightings and the "sense of adventure" in nature, further enhance the overall satisfaction of a visitor to a National Park (Crilley et al., 2012).

In the study published in 2010 on visitor satisfaction in the Dadia-Lefkimmi-Soufli National Forest Park, the criteria were selected partly based on the available literature and partly based on a pilot study that had been carried out to define them precisely. Ultimately, the main criteria used were: (1) staff, (2) the park's physical characteristics, (3) infrastructure, (4) recreational areas and (5) information-communication provided to visitors (Arabatzis & Grigoroudis, 2010).

A study in the Kruger National Park in South Africa showed that the main motivations for tourists visiting national parks include nature itself (Engelbrecht et al., 2015), participation in activities, seeking attractions, special experiences that have not been experienced in the past, a sense of nostalgia for previous experiences, events and memories, escaping everyday life and relaxation (Van der Merwe & Saayman, 2009).

Another study in the Okomu National Park in Nigeria identified nature, confirmation of the expectations that visitors had based on available information about the park, infrastructure and accessibility as the most important criteria that affect visitor satisfaction (Digun-Aweto et al., 2019).

In 2012, a study was conducted on the characteristics-factors that contribute to visitor satisfaction in national parks in the Marmara region of Turkey. The factors used were (a) the natural characteristics of the parks including the landscape, natural beauty, tranquility and historical elements, (b) the quality of services, including information points, museums, and other services provided by the staff, (c) the infrastructure, including signage, the availability of waste bins, toilets, restaurants/cafes and reception areas at the entrances, and (d) the areas used by visitors during their stay in the parks, including resting, observation, picnic and camping areas (Sıvalıoğlu & Berköz, 2012).

It should be noted that the Samaria National Park has some peculiarities in relation to other protected areas and national parks, especially those abroad. As already mentioned, the core of the park is an area of absolute nature protection. Within this area, human activities that are allowed are very limited and mainly concern the "passing" of the Samaria Gorge during the summer months when the park is open. The interventions that can be made in the area to improve the visitor experience are also limited by the same framework.

For the selection of the satisfaction criteria in this study, both the available literature and information provided by the staff of the then park manager, the Chania Forestry Directorate, were taken into account. It was also based on the data that the Chania Forestry Directorate considered useful to study in order to be used in future interventions to achieve even greater visitor satisfaction or to correct possible problems if such a thing was identified by the research.

The criteria that were ultimately selected, as shown in Fig. 4.2, are:

1. **The contribution of tour guides**. Tour guides can significantly influence visitor satisfaction in several ways. First, their knowledge and expertise can provide visitors with a deeper understanding and appreciation of the attractions they visit. A good tour guide can enhance the experience with interesting information, stories and personal touches that visitors would not be able to acquire on their own. In addition, the personality and communication skills of the tour guide play a crucial role. A friendly, enthusiastic and approachable tour guide can make the experience more enjoyable and memorable, while a less communicative or inadequately trained tour guide can reduce visitor satisfaction. Finally, the tour guide's ability to understand both the needs and interests of the group they accompany and to adapt appropriately can ensure that all visitors get the maximum benefit from their visit. Overall, tour guides are a factor that can enhance or diminish visitor satisfaction.

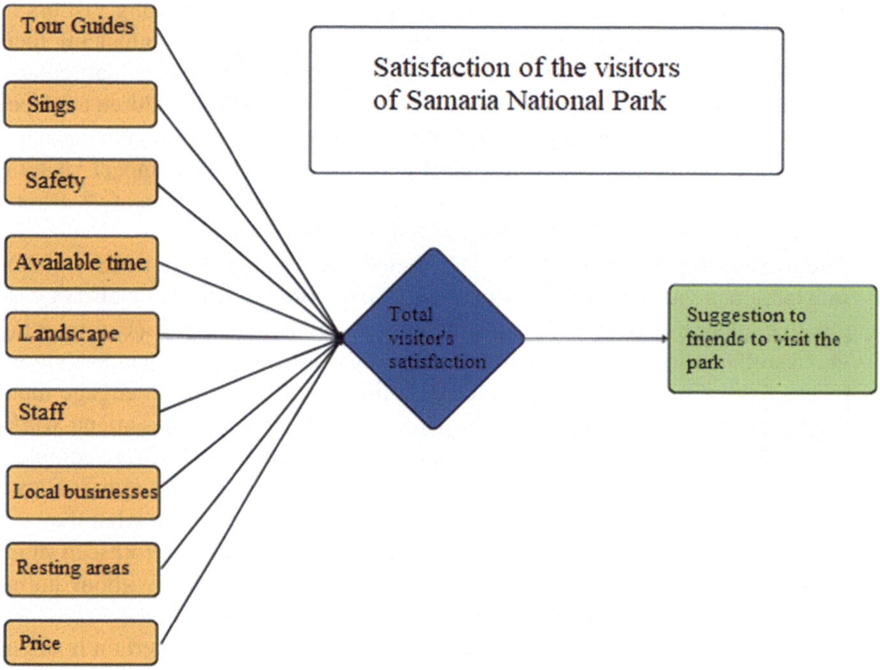

Fig. 4.2 Contribution of criteria to the total satisfaction

2. **The signage along the path**. Clear and well-placed signs help visitors orient themselves and navigate effectively, informational signs about potential hazards, wildlife, and safety protocols ensure that visitors are aware of the dangers and how to stay safe, giving them peace of mind. In addition, signs that provide information about the flora, fauna, geology, and history of the park enrich the visitor experience with educational content, making the visit more interesting. Signs that state park rules help preserve the environment and ensure that all visitors can enjoy the space responsibly. In addition, well-designed signs can enhance the aesthetic experience of the park, contributing positively to satisfaction.
3. **The sense of safety** that visitors feel during their stay in the park. As a factor, it can profoundly influence visitor satisfaction in various settings, such as tourist destinations, recreational areas, and public spaces. When visitors feel safe, they are more likely to relax and enjoy their surroundings, focusing on the experience they can gain rather than worrying about potential dangers. This can encourage them to stay longer and explore more areas, providing a richer and more satisfying experience. A sense of safety contributes to a positive perception of the destination, increasing the likelihood of positive reviews and recommendations to friends and family to visit the destination, attracting more visitors in the future. Furthermore, visitors who feel safe are more likely to return to a place where they previously felt comfortable and enjoyed their experience. Safety also enhances visitor engagement in various activities, leading to a more interactive

and enjoyable experience. Safety provides peace of mind, allowing visitors to focus on learning, exploring and enjoying without distractions, enhancing their overall experience and satisfaction. A safe environment is also particularly important for families, as parents are more likely to bring their children to places where they feel confident about their safety. Finally, effective crisis management and clear communication during emergencies can significantly impact visitors' perceptions of a place. Proper incident handling can mitigate negative experiences and reassure visitors that their well-being is a priority.

4. **The available time in the park**. This factor can significantly influence visitor satisfaction at various tourist destinations. First, more time available allows visitors to explore and enjoy their surroundings at a more leisurely pace, improving the overall quality of their experience. Rushing can lead to stress and less enjoyment of the visit. Second, having enough time allows visitors to engage more deeply with the site and attractions. They can join guided tours, attend workshops, or take part in interactive experiences, which can lead to a more complete and memorable visit. Furthermore, when the available time is enough, it allows visitors to absorb more information, ask questions, and reflect on what they see and learn. Finally, ample time offers flexibility in dealing with unforeseen situations, delays, or changes in plans, ensuring that visitors can adapt without diminishing their enjoyment of the experience.

   Access to the Samaria National Park is only allowed during certain hours of the day and visitors are required to have completed their reading before it closes. Therefore, it was considered important to evaluate the available time as a factor affecting visitor satisfaction.

5. **The landscape**, i.e. all the natural features of the park, including the fauna and flora. A spectacular landscape can provide visitors with a sense of wonder and awe, enhancing the aesthetic enjoyment of the visit. In addition, natural features can offer unique opportunities for activities such as hiking, swimming, camping and photography, making the visit more interactive and enjoyable. The diversity of the landscape and natural features can also offer visitors the opportunity to explore different ecosystems and learn about the biodiversity of the area, which can be of particular interest to nature lovers and scientists. Furthermore, the preservation of the natural environment plays a crucial role in the visitor experience. A well-maintained natural landscape enhances the sense of calm and serenity, contributing to a positive overall experience. Finally, the accessibility of natural landscapes and features can affect visitor satisfaction. Easily accessible points with good infrastructure can enhance the experience, making it more comfortable and enjoyable.

6. **The staff of the Park**. The staff can play an important role in visitor satisfaction. Friendly and professional employees can create a positive first impression and enhance the sense of hospitality. Trained employees who provide accurate and useful information can help visitors make the most of their visit, answering their questions and offering advice on what to see and do. In addition, prompt and efficient service from the staff can resolve any problems or concerns that may arise, improving the overall visitor experience. The presence of employees who are approachable and ready to help at any time enhances visitors' feelings of safety and comfort. Also, employees who are polite and show interest in the

needs of guests can make the experience more personal and enjoyable. Finally, the cleanliness and maintenance of the facilities by the staff contribute to creating a pleasant and healthy environment, which is crucial for guest satisfaction.

7. **Local businesses** located in the wider area. Local businesses, such as taverns, restaurants, hotels and other accommodations, can significantly influence visitor satisfaction in an area. First, the quality of the services offered by these businesses plays a decisive role. A friendly and professional staff, providing prompt and courteous service, can make visitors feel welcome and satisfied. The quality of services and products, such as food in restaurants and cleanliness and comfort in accommodations, is also critical to the visitor experience. Delicious and traditional meals, reflecting local culture, can add value to the visit and make the experience more authentic and memorable. In addition, the variety and availability of options in taverns, restaurants and accommodations can influence visitor satisfaction, providing choices that satisfy different tastes and preferences. A good location of businesses, close to attractions and points of interest, facilitates visitor access and improves their overall experience. Finally, value for money is also important, as visitors want to feel they are getting value for their money.
8. **The rest areas** that have been created within the park. Here, both their location and their infrastructure such as benches, toilets, water sources, etc. are evaluated. Rest areas can significantly influence visitor satisfaction in various environments. The existence of comfortable and clean rest areas offers visitors the opportunity to rest and regain strength, improving their overall experience. These areas can include benches, shade, picnic tables and well-maintained toilets. When visitors can take a break and enjoy their environment in comfortable conditions, this can reduce fatigue and stress. In addition, rest areas that are aesthetically pleasing and well-integrated into the natural environment, as well as properly located, can enhance the aesthetic experience of visitors. The presence of clean and functional facilities, such as trash cans, helps maintain the cleanliness and safety of the area. Rest areas that offer additional amenities, such as drinking water fountains and mobile charging stations, can add value to the visit and make visitors feel more comfortable.
9. **The price of the entrance ticket**. The price can significantly affect visitor satisfaction. Value for money is critical, as visitors want to feel that they are receiving value for their money. If services and amenities meet or exceed expectations in relation to the price they pay, this can increase their satisfaction. Conversely, high prices without corresponding service quality can cause dissatisfaction and disappointment. Competitive and transparent pricing, with no hidden fees, builds guest trust and makes them feel that the cost is fair and justified. Additionally, discounts, offers, and packages that offer extra value can make the experience more attractive and affordable.

In this specific set of questions, the overall satisfaction of the visitors with their visit and their intention to recommend to friends and relatives to visit Samaria National Park in the future was also assessed. For the satisfaction questions, questions with a 5-point Likert scale were used.

References

Arabatzis, G., & Grigoroudis, E. (2010). Visitors' satisfaction, perceptions and gap analysis: The case of Dadia–Lefkimi–Souflion National Park. *Forest Policy and Economics, 12*, 163–172.

Crilley, G., Weber, D., & Taplin, R. (2012). Predicting visitor satisfaction in parks: Comparing the value of personal benefit attainment and service levels in Kakadu National Park, Australia. *Visitor Studies, 15*(2), 217–237.

Digun-Aweto, O., Fawole, O. P., & Van Der Merwe, P. (2019). Nature tourism satisfaction in Okomu National Park, Edo state, Nigeria. *Polish Journal of Sport and Tourism, 26*(4), 32–37.

Engelbrecht, W., Kruger, M., & Saayman, M. (2015). Tourists' view of the factors that gives the Kruger National Park a competitive edge. *Journal of Contemporary Management, 12*, 703–733.

Geng, D. C., Innes, J. L., Wu, W., Wang, W., & Wang, G. (2021). Seasonal variation in visitor satisfaction and its management implications in Banff National Park. *Sustainability, 13*(4), 1681.

McGuire, F. A., O'Leary, J. T., & Dottavio, F. D. (1989). The importance of selected facilities, programs, and services to older visitors to National Parks. *Journal of Park and Recreation Administration, 7*, 1–9.

Silverberg, K. E., Backman, S. J., & Backman, K. F. (1996). A preliminary investigation into the psychographics of nature-based travelers to the Southeastern United States. *Journal of Travel Research, 35*, 19–28.

Sıvalıoğlu, P., & Berköz, L. (2012). User satisfaction in national parks. *Academic Research International, 2*, 537–548.

Tang, X., & Adesina, J. A. (2022). Biodiversity conservation of national parks and nature-protected areas in West Africa: The case of Kainji National Park, Nigeria. *Sustainability, 14*(12), 7322.

Van Der Merwe, P., & Saayman, M. (2009). Travel motivations of tourists visiting Kruger National Park. *Koedoe—African Protected Area Conservation and Science, 50*(1), 154–159.

Zolfaghari, A., & Choi, C. (2023). Elevating the park experience: Exploring asymmetric relationships in visitor satisfaction at Canadian national parks. *Journal of Outdoor Recreation and Tourism, 43*, 100666.

Chapter 5
The Results

5.1 Descriptive Statistics

5.1.1 Visitor Profile

This section presents the demographic characteristics of tourists visiting Samaria National Park. It examines the visitors' gender, education level, age, income, and nationality.

Analyzing the data from the past 5 years (Table 5.1 and Fig. 5.1), a balance is observed in the gender distribution of survey responses. The percentages of male and female visitors vary slightly from year to year but generally remain close to an equilibrium.

Examining the level of education of visitors to Samaria National Park (Table 5.2), we observe the following interesting data:

1. *Primary Education:* Visitors with primary education (Fig. 5.2) are relatively few. However, a notable increase was observed in 2022.
2. *Secondary Education:* A significant portion of visitors reported having completed secondary education (Fig. 5.3).
3. *University Degree:* Many visitors to Samaria National Park hold university degrees (Fig. 5.4). This group consistently constitutes the largest proportion, possibly reflecting a trend for individuals with higher education to choose the park for their excursions.

Table 5.1 Visitor gender distribution

Gender	2018	2019	2020	2021	2022
Male	52.50%	47.80%	48.87%	47.67%	49.10%
Female	47.50%	52.20%	51.13%	52.33%	50.90%

C. Zopounidis, G. Fakotakis, *Multi-Criteria Methodology for the Dynamic Development of Tourist Areas*, Multiple Criteria Decision Making,
https://doi.org/10.1007/978-3-031-95314-9_5

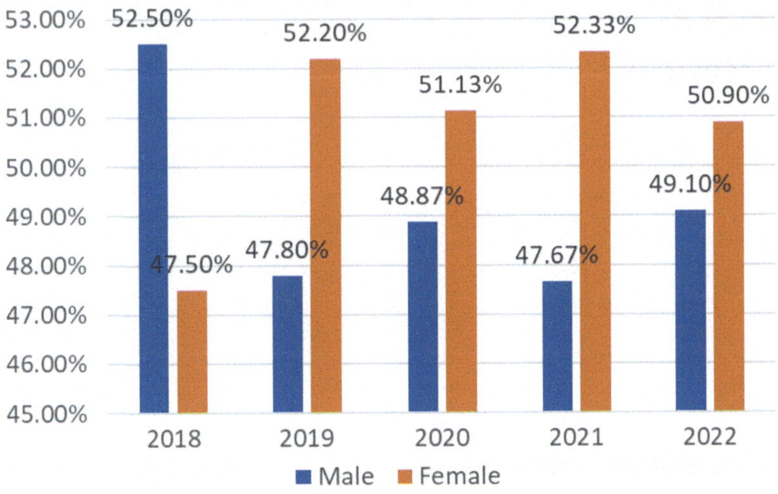

Fig. 5.1 Visitor gender distribution

Table 5.2 Education level of visitors

Education Level	2018	2019	2020	2021	2022
Primary Education	5.90%	5.00%	7.46%	7.30%	17.80%
Secondary Education	27.90%	29.90%	21.69%	30.90%	21.50%
University Degree	34.40%	33.30%	37.29%	39.20%	29.90%
Postgraduate/Doctorate	31.80%	31.80%	33.56%	22.60%	30.80%

4. *Postgraduate/Doctorate:* The number of visitors with postgraduate or doctoral degrees (Fig. 5.5) remains substantial each year, especially compared to lower education levels.

The data portrays a diverse audience visiting Samaria National Park, ranging from individuals with basic education to those with advanced degrees, all drawn to the park's natural beauty.

Examining the data on the age of visitors, as presented in Table 5.3, we observe the following trends:

1. Under 20 Years: This group (Fig. 5.6), despite their enthusiasm for natural beauty, showed a decrease in percentages in 2021. This decline may be due to economic constraints, the pandemic, or shifting preferences and priorities among younger individuals.
2. 20–35 Years: This age group (Fig. 5.7), which constitutes a significant portion of the visitor population, exhibited notable variability. The decline in 2022 may stem from various factors, including economic changes, work commitments, or social events.

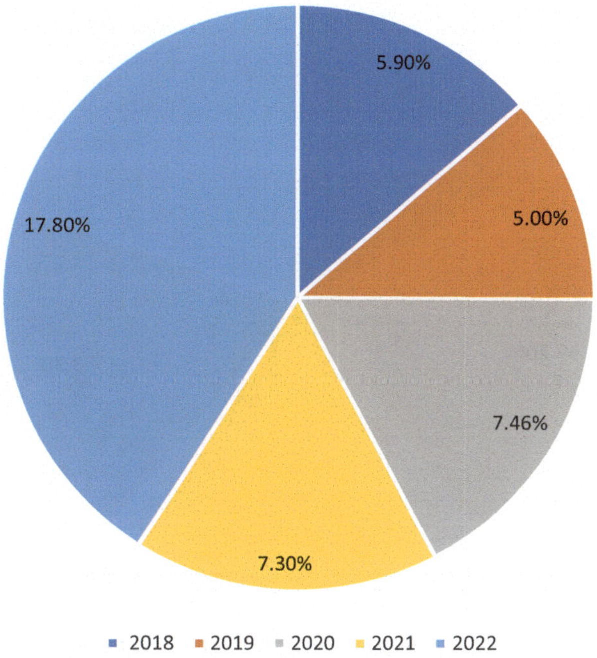

Fig. 5.2 Visitors with primary education

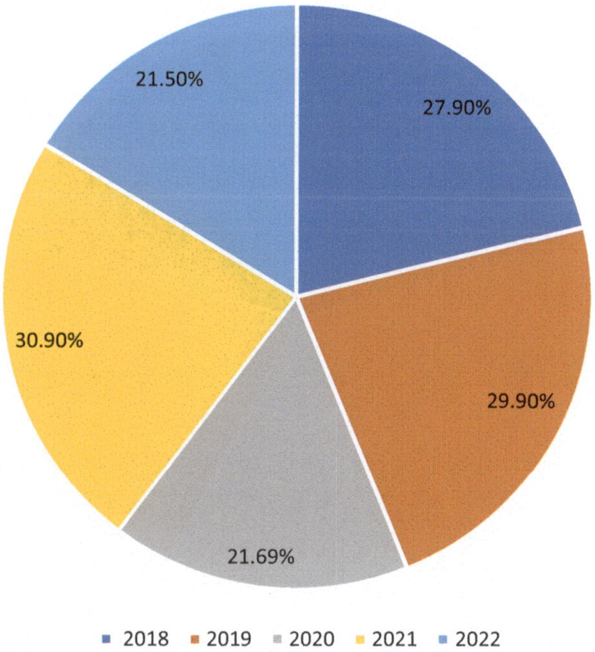

Fig. 5.3 Visitors with secondary education

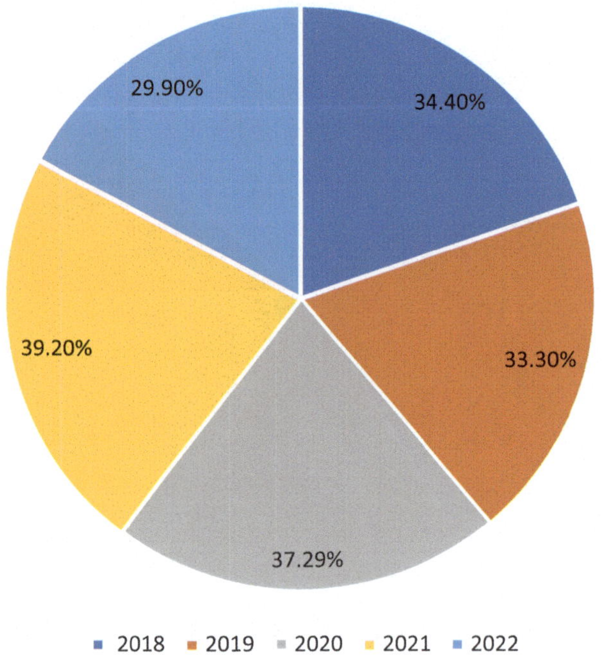

Fig. 5.4 Visitors with a university degree

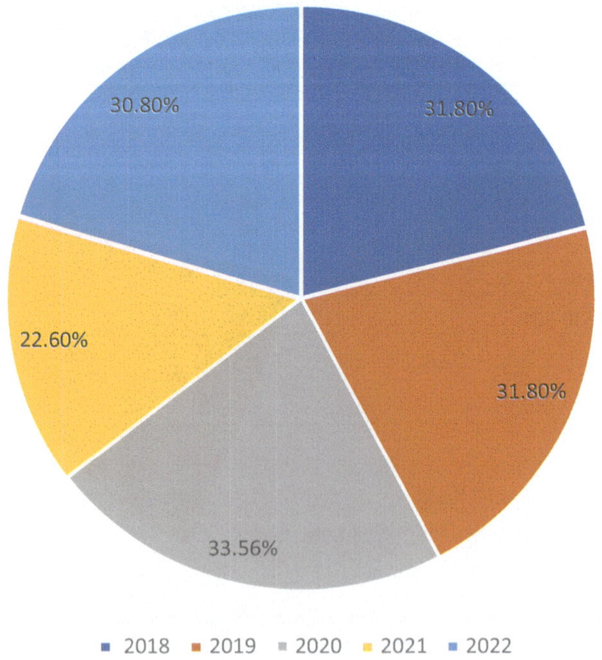

Fig. 5.5 Visitors with postgraduate/doctorate degrees

Table 5.3 Age distribution of visitors

Age group	2018	2019	2020	2021	2022
Under 20 years	13.10%	9.90%	9.97%	4.30%	5.60%
20–35 years	41.90%	43.50%	51.77%	42.60%	30.80%
36–50 years	22.00%	27.90%	21.54%	21.30%	33.60%
51–65 years	18.30%	15.30%	13.83%	24.50%	15.90%
Over 65 years	4.60%	3.40%	2.89%	7.40%	14.00%

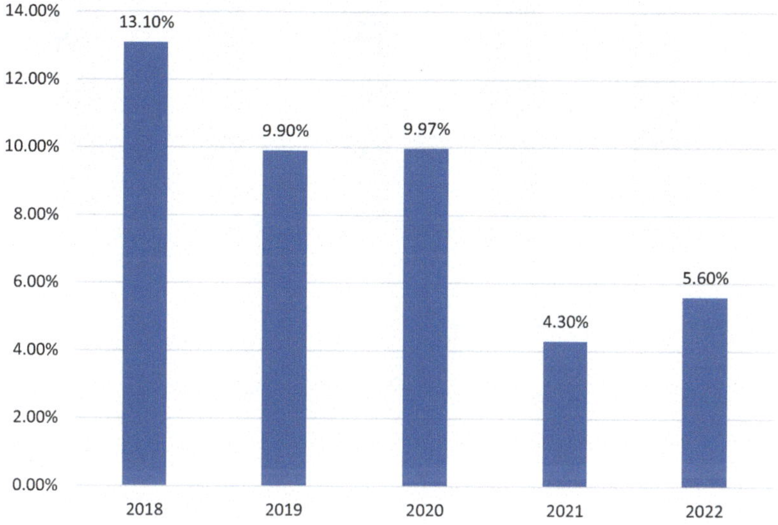

Fig. 5.6 Visitors under 20 years

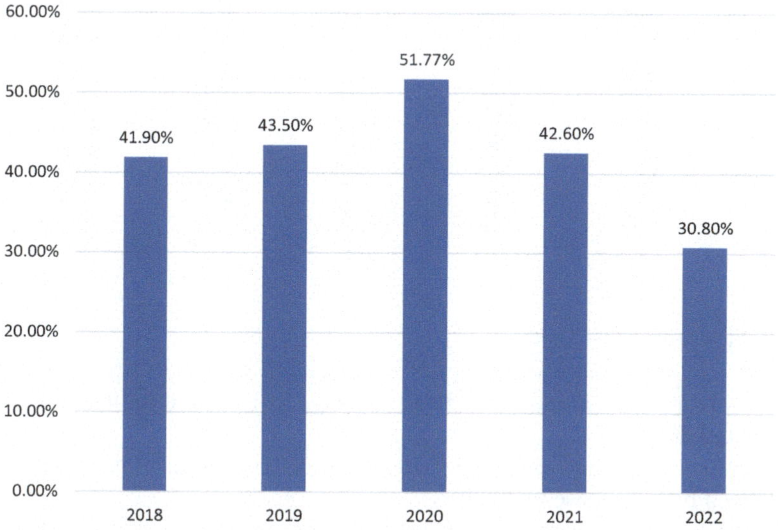

Fig. 5.7 Visitors aged 20–35 years

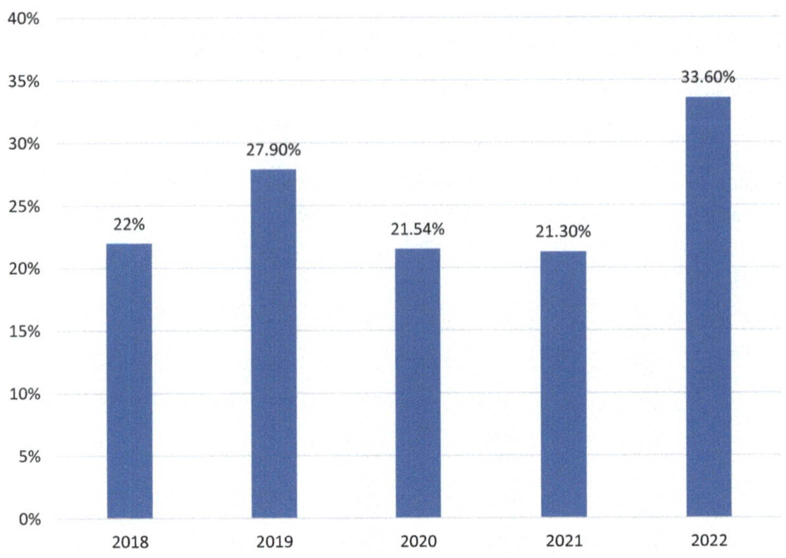

Fig. 5.8 Visitors aged 36–50 years

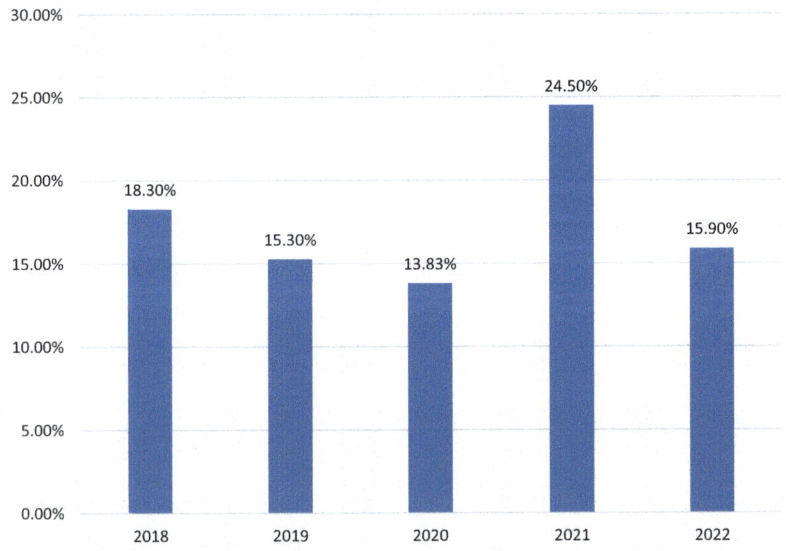

Fig. 5.9 Visitors aged 51–65 years

3. 36–50 Years: Visitors aged 36–50 years (Fig. 5.8) showed an increase in both 2019 and 2022.
4. 51–65 Years: This group (Fig. 5.9) demonstrated a significant increase in 2021, followed by a decrease in 2022.
5. Over 65 Years: The increase in visitors in this age group in recent years (Fig. 5.10) is noteworthy.

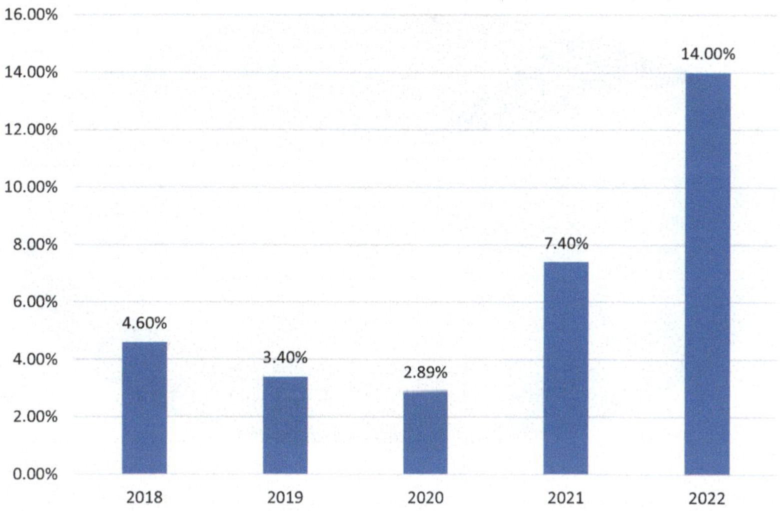

Fig. 5.10 Visitors over 65 years

Table 5.4 Annual income of visitors

Annual income (€)	2018	2019	2020	2021	2022
Up to 10,000	17.20%	17.90%	32.35%	18.60%	19.10%
10,001–18,000	8.10%	9.50%	14.34%	12.40%	20.20%
18,001–26,000	15.10%	16.50%	9.56%	4.10%	19.30%
26,001–34,000	13.00%	9.50%	4.41%	9.30%	10.80%
34,001–45,000	21.10%	18.80%	15.44%	5.20%	12.00%
45,001–60,000	8.10%	12.10%	7.35%	10.30%	7.40%
60,001–80,000	8.10%	8.10%	7.35%	15.50%	6.90%
Over 80,000	9.30%	7.40%	9.19%	10.30%	4.20%

Overall, Samaria National Park continues to attract visitors across all age groups. The majority of visitors fall within the 20–50 age range, reflecting the nature of the park. Traversing the park requires hiking a 13-kilometer trail, necessitating a certain level of physical fitness. However, in recent years, post-pandemic trends indicate an increase in visitors over 65 years of age.

Observations by Year (Table 5.4):

- 2018: The majority of visitors fell into the 34,001–45,000 € (21.10%) and 18,001–26,000 € (15.10%) categories (Fig. 5.11). A relatively lower percentage of visitors reported incomes over 80,000 € (9.30%).
- 2019: Visitor incomes showed a similar yet slightly more balanced distribution among income brackets (Fig. 5.12). The highest percentage was in the 18,001–26,000 € category (16.50%).
- 2020: In the first year of the pandemic (Fig. 5.13), there was a significant increase in visitors with incomes up to 10,000 € (32.35%). This likely reflects the economic impact of lockdowns and changes in visitor demographics, as explored further below.

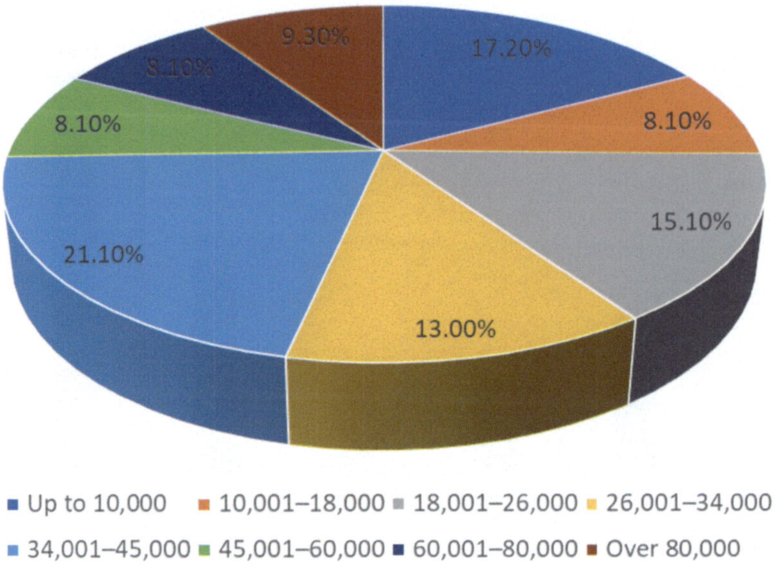

Fig. 5.11 Annual income 2018

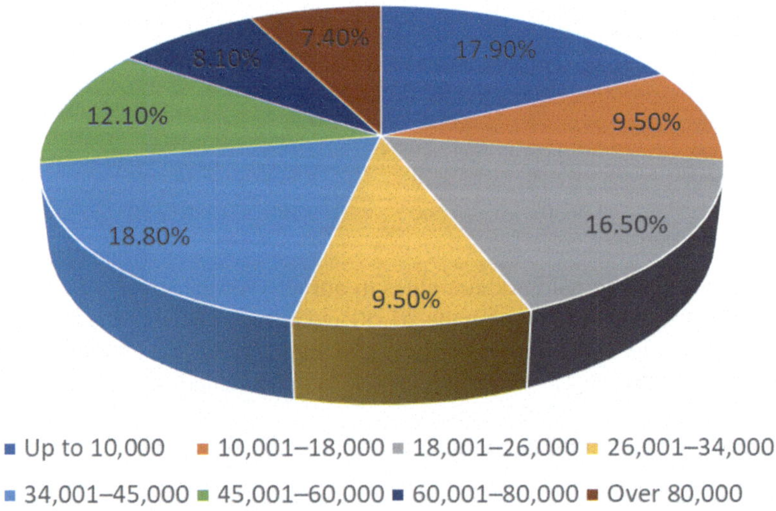

Fig. 5.12 Annual income 2019

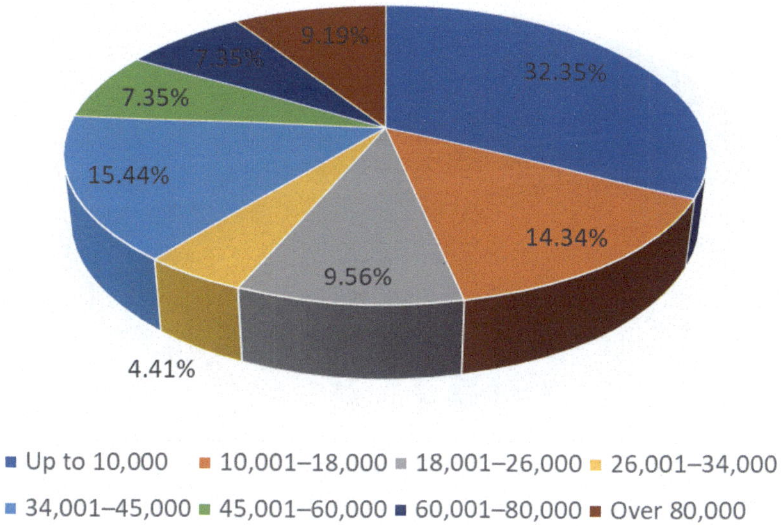

Fig. 5.13 Annual income 2020

- 2021: In contrast to 2020, 2021 (Fig. 5.14) saw a return closer to pre-pandemic income distributions. Notable changes included a decline in the 18,001–26,000 € category (4.10%) and an increase in the 60,001–80,000 € group (15.50%).
- 2022: The final year of the study (Fig. 5.15) revealed significant shifts in income brackets, including increases in the 10,001–18,000 € (20.20%) and 18,001–26,000 € (19.30%) groups. Conversely, incomes over 80,000 € declined sharply to 4.20%.

Trends in visitor income demonstrate an intriguing ebb and flow across years, highlighting the diverse economic backgrounds of Samaria National Park's visitors. This diversity underscores the park's appeal to a wide range of income levels.

5.1.2 Visitor Origin

2020 was a special year for the nationality of visitors (Table 5.5 and Fig. 5.16) in Samaria National Park. While in previous years the vast majority of visitors came from abroad, in 2020 this trend changed considerably. Specifically, the percentage of Greek visitors increased significantly, reaching 43.6%, almost quadrupling the percentages of previous years. In contrast, visitors from abroad amounted to 56.31%, reducing their dominance in the total number of visitors that they showed in previous years.

A change in the origin of visitors between years could generally be attributed to many factors, but the most obvious here is the COVID-19 pandemic. Due to travel restrictions and closed borders in many countries, but also fear, traveling abroad has

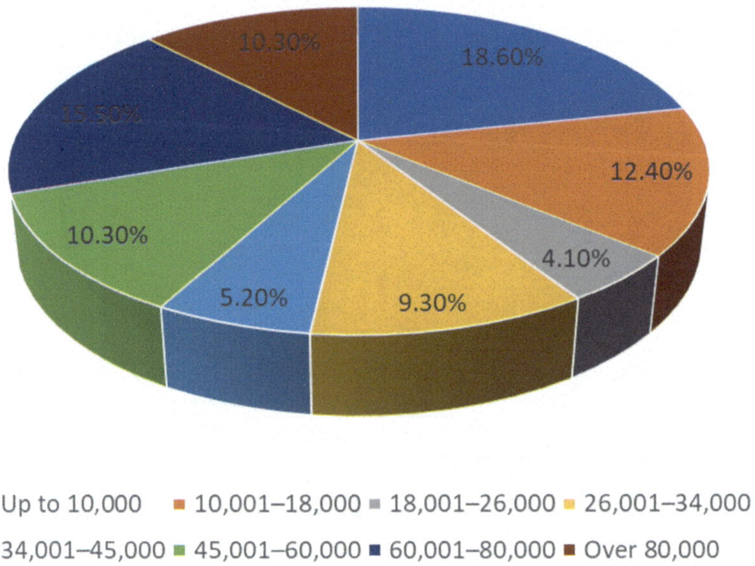

Fig. 5.14 Annual income 2021

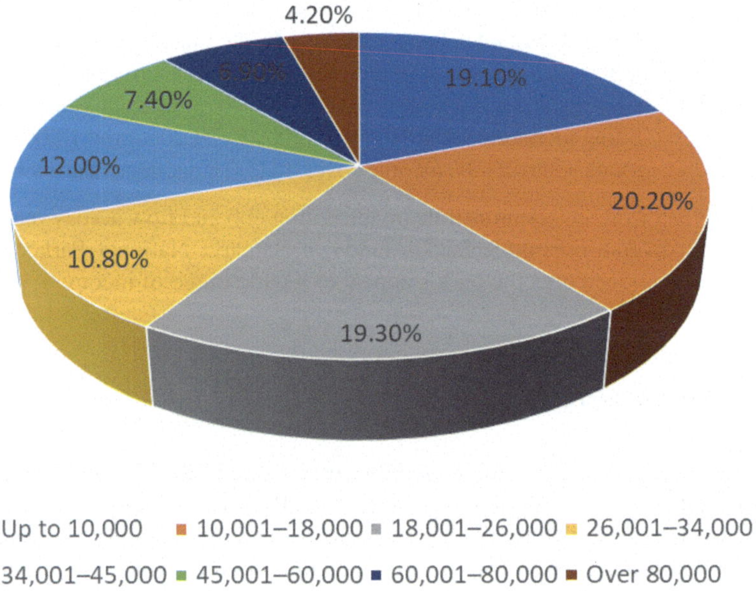

Fig. 5.15 Annual income 2022

Table 5.5 Visitor origin

Origin	2018	2019	2020	2021	2022
Greece	13.90%	11.90%	43.60%	14.50%	14.10%
Abroad	86.10%	88.10%	56.31%	85.50%	85.90%

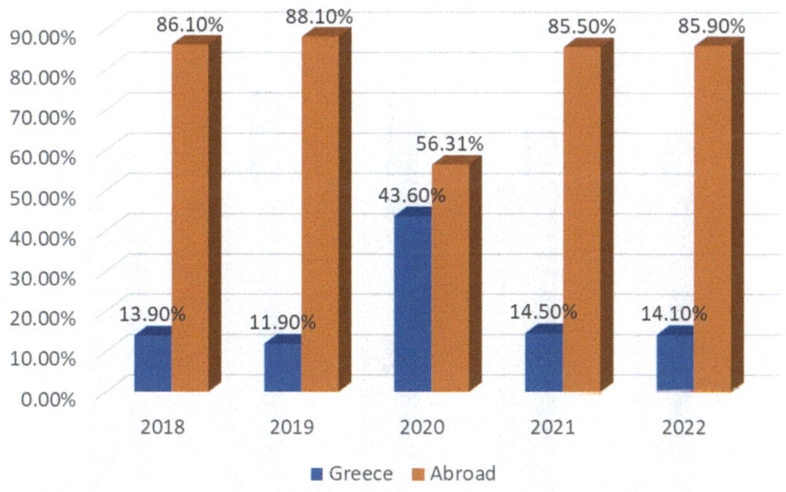

Fig. 5.16 Visitor origin

Table 5.6 Information sources about Samaria National Park before arriving in Chania

Source	2018	2019	2020	2021	2022
From the Internet	46.80%	47.80%	50.70%	52.10%	45.10%
From friends and relatives who visited before	42.60%	41.70%	38.10%	41.50%	45.10%
No information	4.70%	4.30%	4.50%	3.20%	3.90%
From other sources	5.90%	6.20%	6.70%	3.20%	5.90%

become difficult for many. On the other hand, many Greeks have chosen to spend their holidays within the country, discovering the natural beauty in a foreign and virus-safe place, and Samaria National Park has been an ideal outlet.

5.1.3 Characteristics of the Visit

This section presents information regarding visits to Samaria National Park, including the sources through which visitors learn about the park, the role the park played in tourists' decision to choose Crete for their holidays, the means of transportation used to reach the park, and whether it was their first visit or a return trip.

As shown in Table 5.6 and Fig. 5.17, over the 5 years of the survey, from 2018 to 2022, patterns emerged regarding how visitors obtained information about Samaria National Park before their visit to Chania. It is worth noting that the Internet consistently stood out as the main source. Alongside this, recommendations from friends and relatives who had visited the park in the past proved to be a consistent source of information. On the other hand, it is interesting to note that a small percentage of visitors, on average around 4%, were recorded who visited the park without knowing its existence before arriving in Crete. Finally, there are some alternative sources of information for a small percentage of visitors such as books and relevant information leaflets.

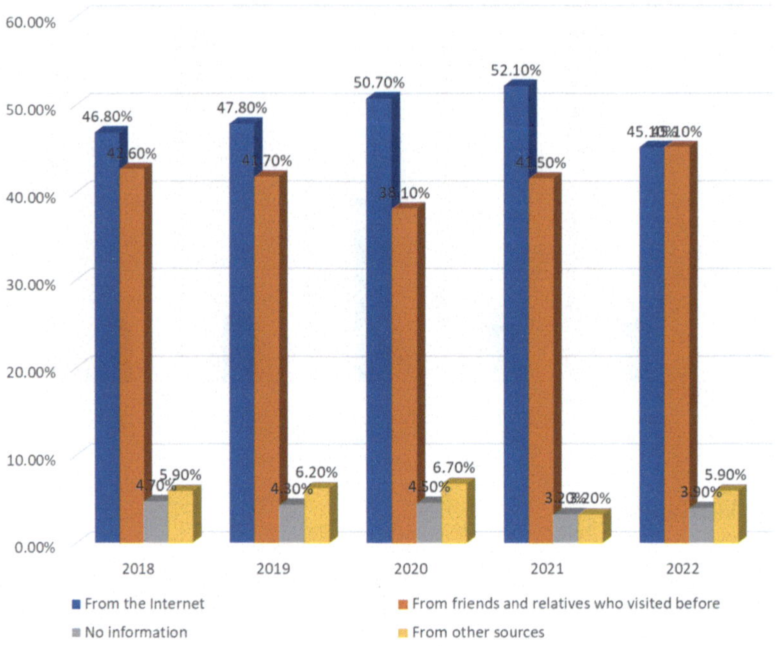

Fig. 5.17 Sources of information about the park

Table 5.7 Means of transportation to and from the park

Year	2018	2019	2020	2021	2022
Public Bus (KTEL)	22.30%	23.40%	32.46%	10.40%	20.60%
Tour Agency Bus	57.10%	55.70%	42.95%	51.00%	42.10%
Car	14.90%	16.40%	20.33%	32.30%	29.00%
Other	5.70%	4.50%	4.26%	6.30%	8.40%

In essence, while digital platforms and personal recommendations dominated as sources of information about the park, other sources and visits without prior knowledge continued to play a role, albeit to a much lesser extent.

As shown in Table 5.7 and Fig. 5.18, tour agency buses remained the most popular mode of transportation to and from the park throughout the years. In 2018, 57.10% of visitors used this option, although it slightly decreased to 55.70% in 2019. However, in 2020, there was a significant decline, with only 42.95% of visitors opting for this mode. This sharp drop during the first year of the pandemic is largely attributed to changes in the origin of visitors and the preference of Greek visitors for public buses (KTEL). The popularity of tour agency buses rebounded to 51% in 2021 but declined again in 2022 to 42.10%.

At the same time, the use of public buses (KTEL) (Fig. 5.19) showed fluctuations. Starting at 22.30% in 2018, it surged to 32.46% in 2020, for the reasons mentioned earlier. However, this percentage dropped significantly to 10.40% in 2021 as the pandemic's impact lessened, and rose again to 20.60% in 2022.

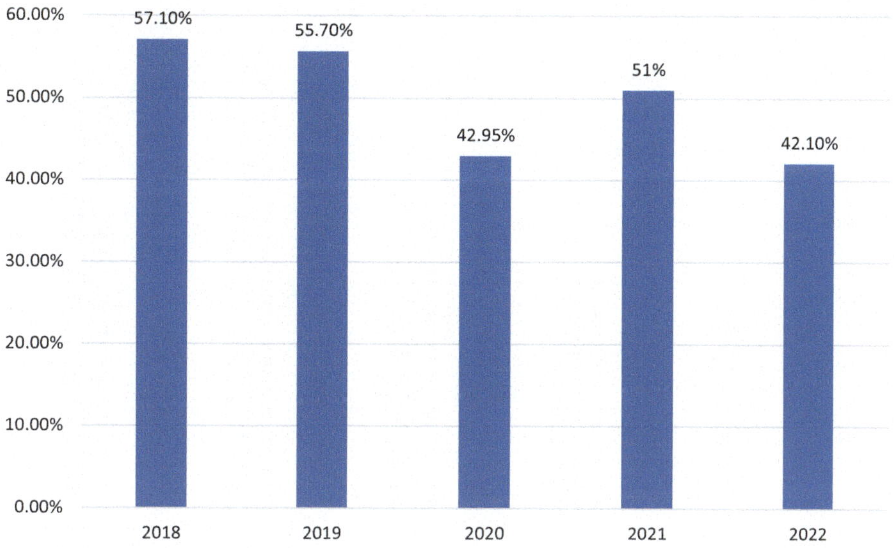

Fig. 5.18 Use of tour agency buses

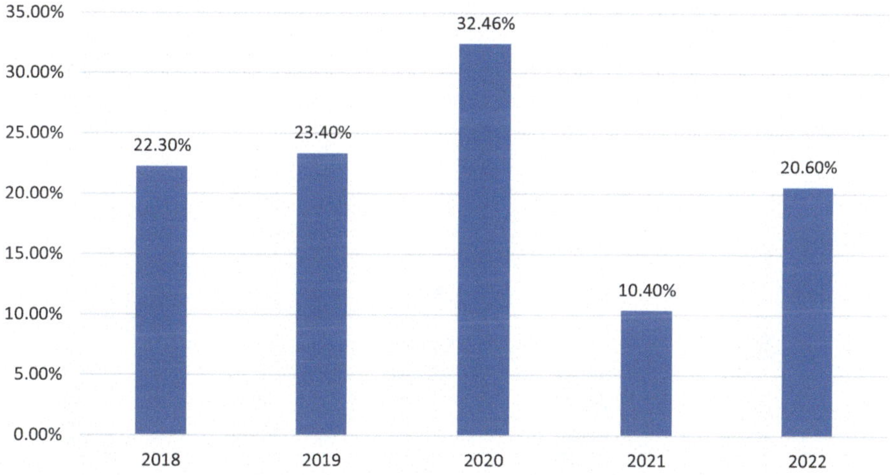

Fig. 5.19 Use of public buses (KTEL)

The use of cars as a mode of transportation (Fig. 5.20) became increasingly popular over the years. Starting at 14.90% in 2018, it saw a significant rise to 32.30% in 2021. By 2022, cars maintained a substantial share, with 29% of visitors driving themselves to the park.

Lastly, other means of transportation were used by a very small proportion of visitors, as illustrated in Fig. 5.21.

While tourist agency buses remained the dominant choice, there is a noticeable shift toward more personalized transportation options, as evidenced by the increasing preference for cars and other means of transport.

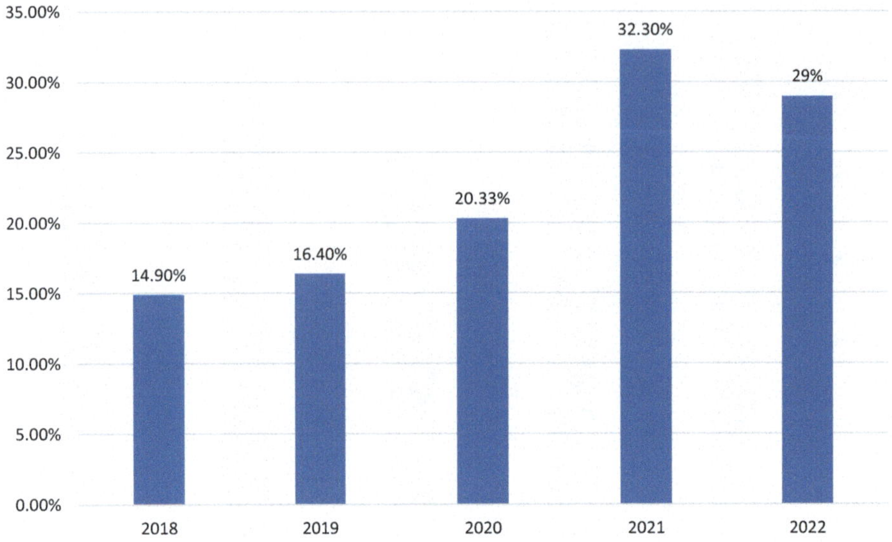

Fig. 5.20 Use of personal cars

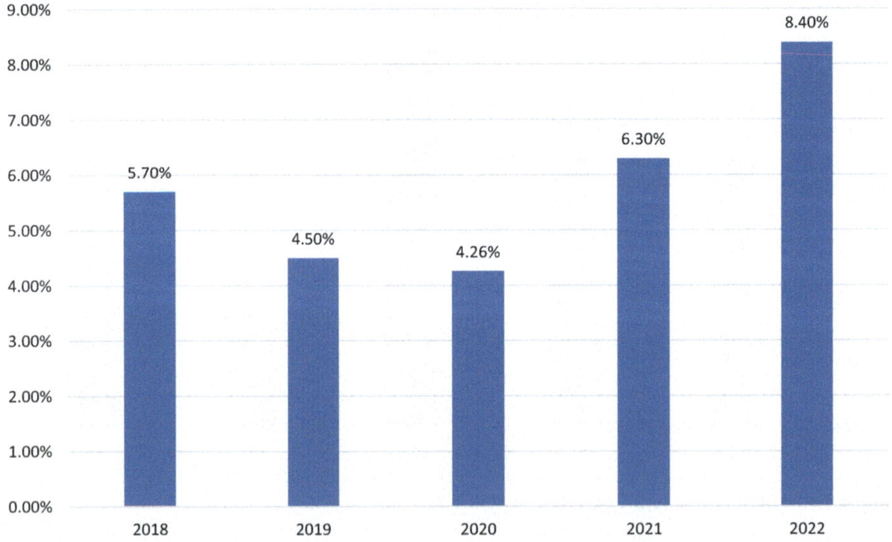

Fig. 5.21 Use of other means

The Table 5.8 and Fig. 5.22 illustrate the percentage of visitors who were visiting the Samaria National Park for the first time versus those who had visited it before.

In 2018, a significant majority (90.20%) reported that it was their first visit to the park. This number remained relatively high in 2019 at 88.60%. However, in 2020, there was a notable shift, with first-time visitors accounting for only 60.65%. This

Table 5.8 Is this your first visit?

Year	2018	2019	2020	2021	2022
Yes	90.20%	88.60%	60.65%	75.80%	67.30%
No	9.80%	11.40%	39.35%	24.20%	32.70%

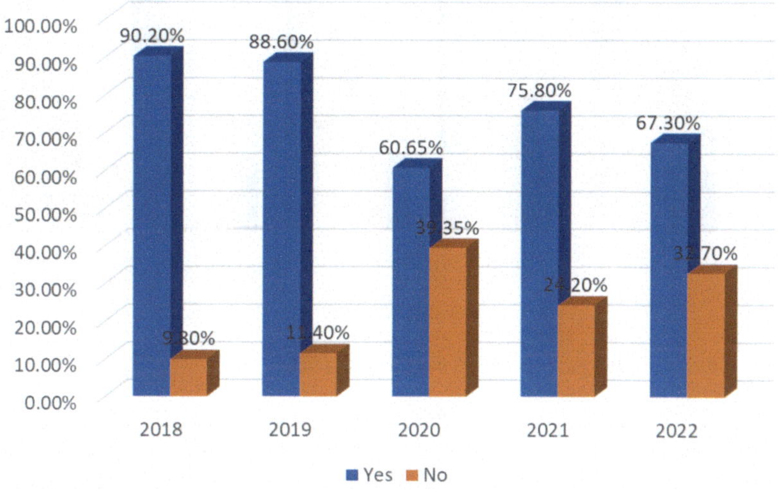

Fig. 5.22 Is it your first visit?

Table 5.9 How much did the existence of Samaria National Park influence your decision to visit Crete?

	2018	2019	2020	2021	2022
5	11.50%	10.80%	8.47%	8.20%	4.60%
4	16.90%	15.50%	11.64%	13.60%	13.40%
3	19.30%	20.90%	23.28%	21.50%	20.20%
2	12.50%	12.40%	22.75%	25.80%	32.60%
1	39.80%	40.30%	33.86%	30.90%	29.20%

drop can be attributed to the pandemic's impact, as a large percentage of the visitors were Greek nationals who had visited the park previously. In 2021, the percentage rose to 75.80%, but it decreased again in 2022 to 67.30%.

On the other hand, the percentage of repeat visitors increased significantly. In 2018, only 9.80% of visitors had visited the park before. This number rose slightly to 11.40% in 2019. A substantial increase occurred in 2020, with 39.35% of visitors reporting they had been to the park previously. This figure dropped to 24.20% in 2021 but rose again in 2022 to 32.70%.

The Table 5.9 and Fig. 5.23 present the role that Samaria National Park played in visitors' decision to travel to Crete between 2018 and 2022. Ratings ranged from 5 (indicating a significant influence) to 1 (indicating no influence at all).

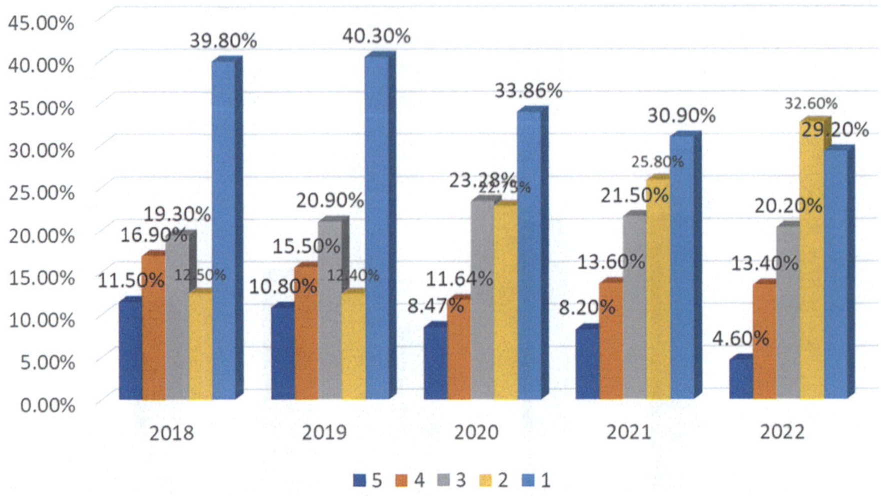

Fig. 5.23 Influence of the park on the decision to visit Crete

In 2018, the highest percentage of visitors (39.80%) indicated that the national park had no influence (rating 1) on their decision to visit Crete. This percentage remained similar in 2019 at 40.30%. However, this percentage began to decline in subsequent years, dropping to 33.86% in 2020, 30.90% in 2021, and 29.20% in 2022. This trend suggests that, over the years, an increasing number of visitors considered the park as having some influence on their travel decision, even if it wasn't the main factor.

At the higher end of the scale, in 2018, 11.50% of respondents gave the maximum rating of **5**, indicating that the park was the primary reason for their visit. This percentage slightly declined in subsequent years, reaching 4.60% in 2022. Similarly, the percentage of visitors who rated the influence as 4 decreased from 16.90% in 2018 to 13.40% in 2022.

Interestingly, there was a noticeable shift in the middle ratings over the years. For rating 3, the percentage of respondents rose from 19.30% in 2018 to 20.20% in 2022. Rating 2, which indicates slight influence, saw the most significant increase, jumping from 12.50% in 2018 to 32.60% in 2022.

While a significant portion of visitors still believed that the park had little to no influence on their decision to visit Crete, there has been a clear trend over the five years showing an increasing number of visitors recognizing that the park played some role in their choice.

Table 5.10 What would you do if the ticket for Samaria National Park was higher?

	2018	2019	2020	2021	2022
I would not visit	17.20%	17.50%	14.08%	10.30%	15.10%
I would pay 2€ more	27.60%	21.80%	33.21%	28.90%	29.20%
I would pay 5€ more	34.10%	40.10%	29.96%	38.10%	34.90%
I would pay 10€ more	21.10%	20.60%	22.74%	22.70%	20.80%

5.1.4 *Evaluation of the Visit*

In this section we examine the evaluations of park visitors for various elements of the visit, for the satisfaction of visitors with their visit and we also see their willingness to pay a higher price for the ticket (Willingness to pay) in order to enter the park (Table 5.10).

In recent years, the concept of "willingness to pay" (WTP)—the maximum amount that intended to be parted with by the consumer for a particular product, service or experience—has contributed to understanding the value that consumers attach to the experiences they gain. For natural attractions such as Samaria National Park, visitors' WTP is derived from their willingness to pay a certain amount for access to it and is evidence of the value and importance of the park.

As shown in Fig. 5.24, in 2018 for 17.20% of visitors, the valuation of the park experience was closely linked to the existing ticket price. However, this percentage is small. A significant 27.60% of visitors would be willing to pay up to €2 more. Even more remarkable, 34.10% were willing to pay up to €5 more, indicating a strong correlation with the value they perceive the park to have for them. A further 21.10% expressed that they might go as far as paying €10 more.

Moving on to 2019 (Fig. 5.25), 17.50% of visitors felt that the current pricing was their limit, while 40.10% of visitors were willing to accept a €5 increase, revealing a growing appreciation for the park. Correspondingly, there was a drop in those who would pay just €2 more, to 21.80%, while the highest category remained relatively stable.

In 2020 (Fig. 5.26) there were changes probably due to external influences and in particular the consequences of the pandemic. The percentage of visitors who felt discouraged by a possible price increase decreased to 14.08%. The willingness to pay an extra €2 or €5 shifted to 33.21% and 29.96% respectively, with those willing to contribute up to €10 more noting a small increase.

In 2021, (Fig. 5.27) the data showed that only 10.30% of visitors expressed reservations about a price increase. The willingness to pay €2, €5 and €10 more categories showed percentages of 28.90%, 38.10% and 22.70% respectively.

The year 2022 (Fig. 5.28) recorded another reversal. 15.10% of visitors do not seem willing to pay more money to access the park. The remaining categories (€2, €5 and €10 more) showed percentages of 29.20%, 34.90% and 20.80%, showing that the value of the park remained high for most visitors.

Table 5.11 and Fig. 5.29 present the average ratings of visitors for various elements of their visit over the years.

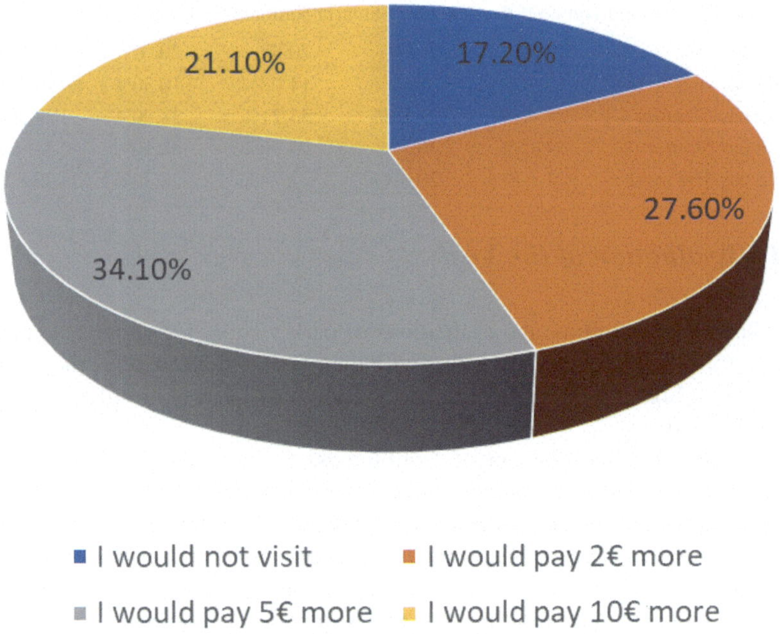

Fig. 5.24 Visitors 2018: What would you do if the park ticket was higher?

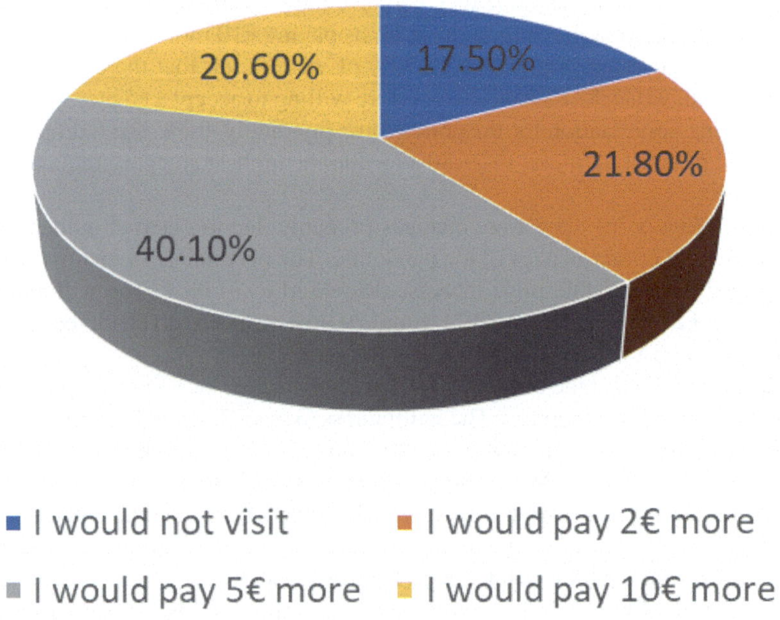

Fig. 5.25 Visitors 2019: What would you do if the park ticket was higher?

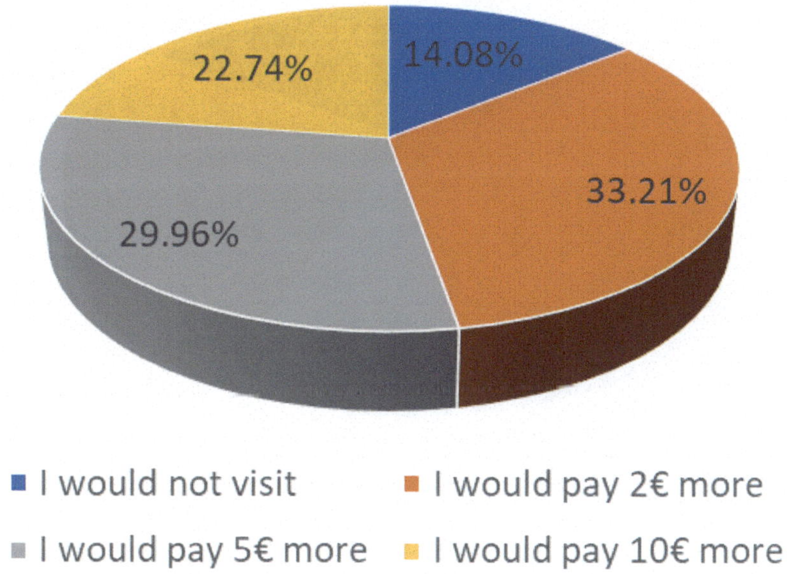

Fig. 5.26 Visitors 2020: What would you do if the park ticket was higher?

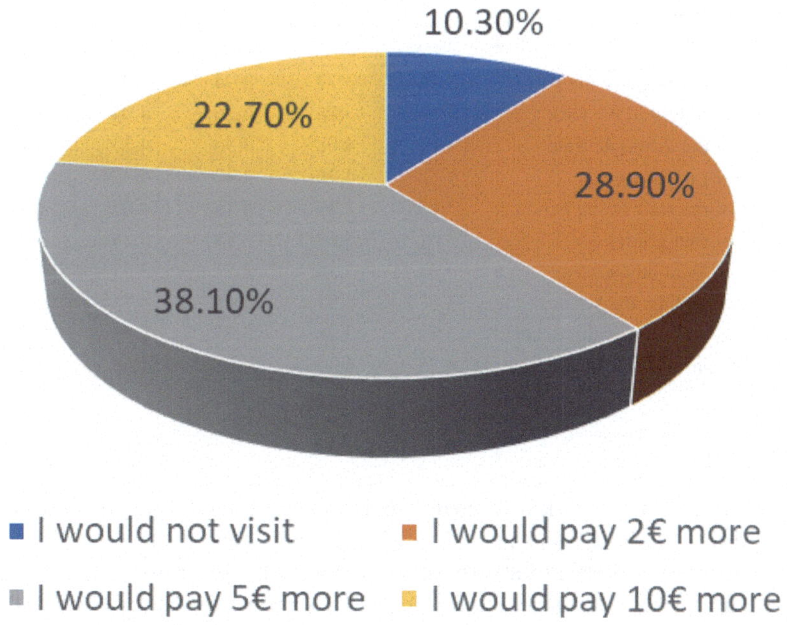

Fig. 5.27 Visitors 2021: What would you do if the park ticket was higher?

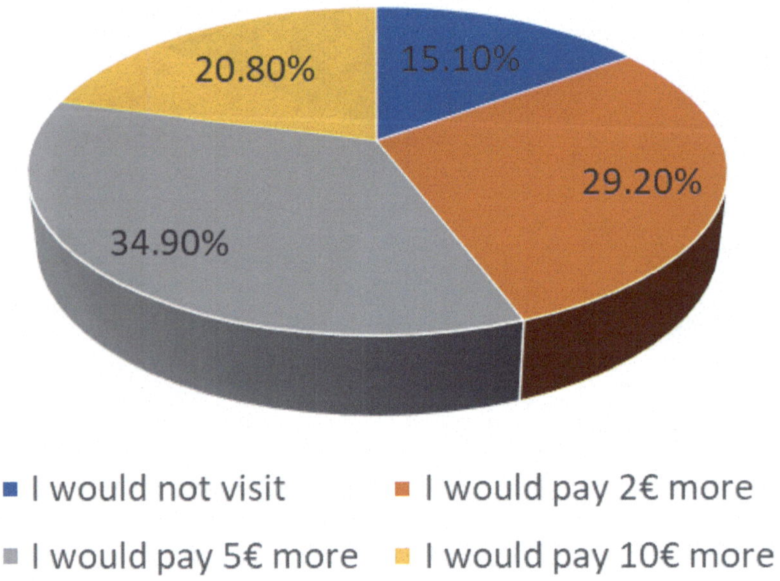

Fig. 5.28 Visitors 2022: What would you do if the park ticket was higher?

Table 5.11 Visit rating

	2018	2019	2020	2021	2022
Evaluation of Signs along the path	3.76	3.72	3.94	3.96	3.97
Evaluation of Safety in the park	3.86	3.88	4.26	4.04	3.612
Evaluation of the available time	4.13	4.08	4.36	4.33	3.72
Evaluation of Landscape	4.32	4.36	4.42	4.37	4.12
Evaluation of the staff of the park	3.85	3.89	4.15	4.07	4.06
Evaluation of tour guides	3.6	3.64	3.78	3.88	3.69
Evaluation of local businesses	3.34	3.36	3.7	3.57	4.14
Evaluation of resting areas	3.87	3.91	4.41	4.15	4.17
Evaluation of the price of the ticket	3.7	3.76	3.99	3.77	3.73
Total evaluation	4.35	4.37	4.61	4.53	4.23

In 2018 (Fig. 5.30):

Signage throughout the park was rated 3.76, indicating a good, though not excellent, rating for the various signs posted by the park managers.

The feeling of safety within the park scored a commendable 3.86.

The time available received a high rating of 4.13, indicating that visitors felt they had enough time to explore and enjoy the park.

The park's landscape also received a high rating with a score of 4.32.

The park staff received an encouraging rating of 3.85, reflecting their competence and good interaction with visitors.

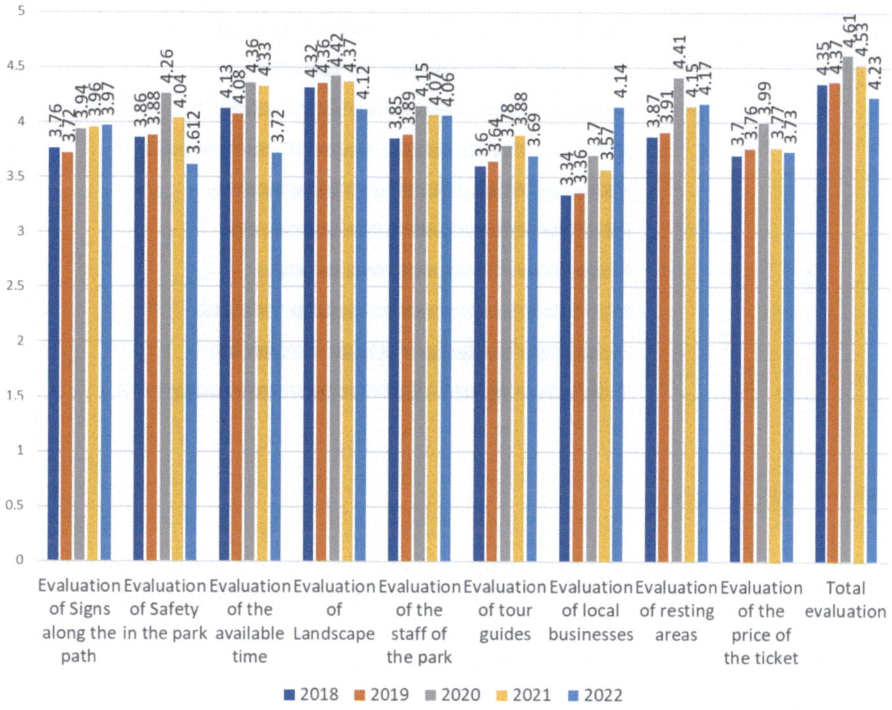

Fig. 5.29 Visit rating by year

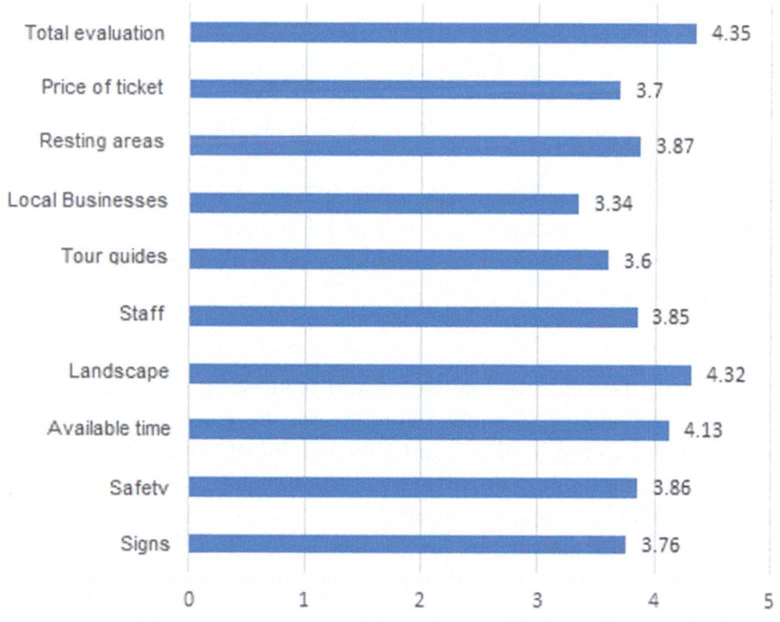

Fig. 5.30 2018 visit rating

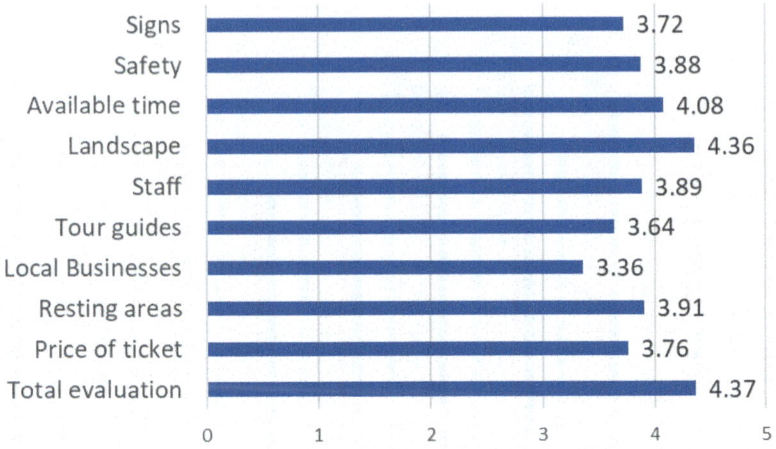

Fig. 5.31 2019 visit rating

The tour guides were rated 3.6, a slightly lower rating but still in the positive range.

Local businesses around the park were rated 3.34, indicating that there was room for improvement in their services or products.

Rest areas received a score of 3.87, indicating their adequacy and comfort.

Ticket pricing was considered fair, with a score of 3.7.

Visitors' overall experience of the park was rated 4.35.

In 2019 (Fig. 5.31):

Most categories showed marginal fluctuations. Signage was rated slightly lower at 3.72, while safety reached 3.88.

Ratings for time available and scenery were relatively stable at 4.08 and 4.36, respectively.

There was a slight increase in ratings for park staff and tour guides, with scores of 3.89 and 3.64.

Local businesses maintained their same score at 3.36.

Rest areas and ticket prices were rated 3.91 and 3.76, respectively.

The overall experience remained consistently good, with a slight improvement in the rating to 4.37.

In 2020 (Fig. 5.32):

The first year of the pandemic saw an overall increase in ratings, with the most notable change being the safety rating within the park, which rose to 4.26. Visitors affected by the conditions after the first lockdown seem to have particularly appreciated the safety provided by a visit to nature.

The rating of available time and scenery received scores of 4.36 and 4.42, suggesting increased visitor satisfaction here as well.

Staff, guides and local businesses all had increased ratings, with scores of 4.15, 3.78 and 3.7.

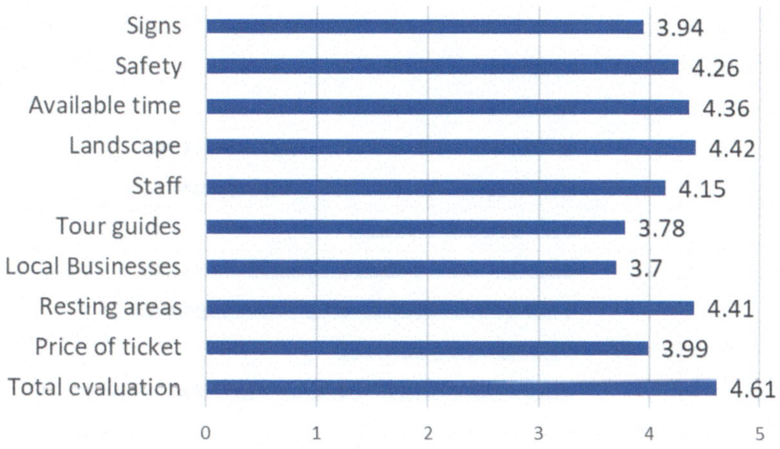

Fig. 5.32 2020 visit rating

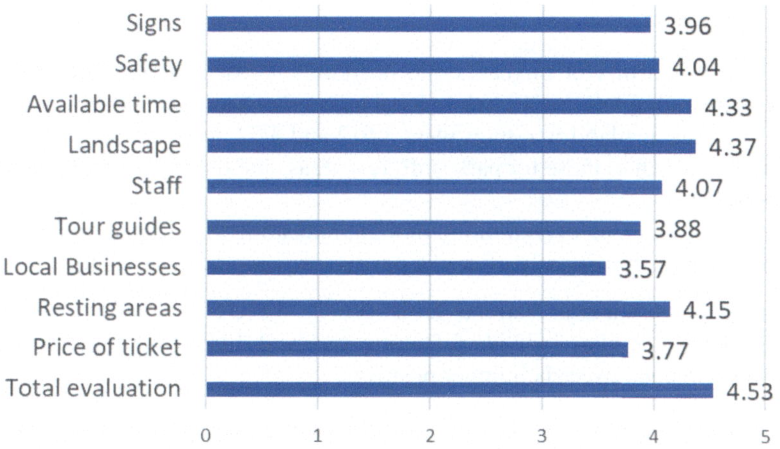

Fig. 5.33 2021 visit rating

Rest areas and ticket prices also saw increases to 4.41 and 3.99.
The year ended with an impressive overall experience score of 4.61.

In 2021 (Fig. 5.33):

Visitors again appeared highly satisfied with little variation in ratings. Safety and time availability were rated 4.04 and 4.33.
The scenery, park staff and guides had scores of 4.37, 4.07 and 3.88.
Local businesses and rest areas received ratings of 3.57 and 4.15, while the ticket price rating was 3.77.
The overall rating was 4.53, slightly lower than the previous year, but still quite high.

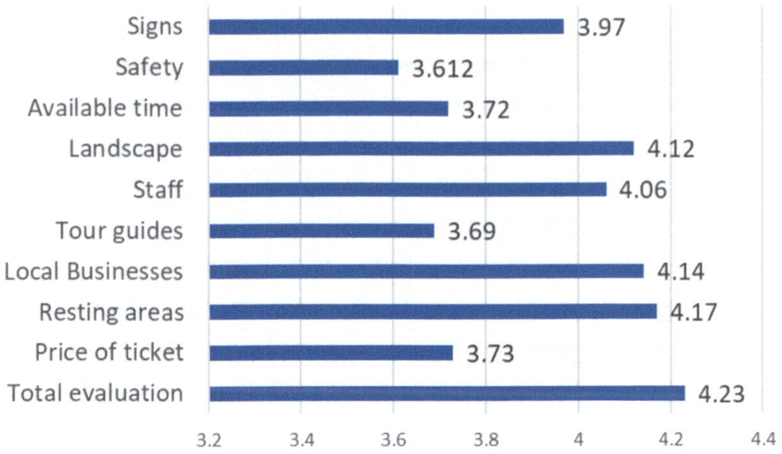

Fig. 5.34 2022 visit evaluation

In 2022 (Fig. 5.34):

Here there was a mix of increases and decreases across different attributes. The landscape rating fell to 4.12, while local businesses increased to 4.14.

Safety and time availability ratings were 3.612 and 3.72, respectively.

Other ratings, such as those for park staff, rest areas, and ticket pricing, ranged from 3.69 to 4.17.

The year ended with an overall rating of 4.23, indicating that despite minor fluctuations, the park consistently offered a positive experience to its visitors.

Over the 5 years, Samaria National Park seems to have consistently impressed its visitors, and while some individual features received different ratings, the overall sentiment remained overwhelmingly positive.

5.2 Application of the MUSA method

The results obtained by applying the MUSA method to the data include the calculation of fit and stability indices, satisfaction functions, satisfaction indices accompanied by the corresponding weights, demand indices and efficiency indices for each year of the survey.

5.2.1 Fit and Stability Indices

As we see in Table 5.12, the fit (AFI) and stability (ASI) indices of the model appear to be good for all the years in which the survey was conducted.

Table 5.12 Adaptation and stability indices

	AFI1	AFI2	AFI3	ASI
2018	0.895	0.089	0.883	0.749
2019	0.886	0.072	0.873	0.729
2020	0.878	0.073	0.868	0.791
2021	0.907	0.034	0.896	0.812
2022	0.894	0.138	0.877	0.692

Table 5.13 MUSA satisfaction results 2018–2019

	2018		2019	
	Level of satisfaction	Weight	Level of satisfaction	Weight
Total satisfaction	0.891		0.885	
Tour guides	0.758	10.33	0.721	8.67
Signs	0.76	9.22	0.746	9.01
Safety	0.869	15.88	0.901	21.08
Available time	0.91	17.92	0.846	10.89
Landscape	0.889	11.11	0.896	11.11
Staff	0.771	8.96	0.817	10.89
Local Businesses	0.629	7.93	0.66	8.60
Resting areas	0.805	10.43	0.79	9.33
Price	0.72	8.21	0.785	10.42

5.2.2 *MUSA Satisfaction Results*

Tables 5.13 and 5.14 present the overall results of the MUSA method for visitor satisfaction for the years before the pandemic and for the years after the pandemic respectively. Specifically, we see the overall level of satisfaction per year, the level of satisfaction from each criterion per year, and the weight of each criterion per year.

Visitors as shown in Fig. 5.35 are quite satisfied with their experience. Overall satisfaction scores are quite high, namely: 89.1% for the 2018 season, 88.5% for the 2019 season, 91.6% for 2020 which was also the first year of the pandemic, 89.3% for 2021 which also had a significant impact on tourism and 83.8% for 2022, the first year without pandemic restrictions.

5.2.3 *Satisfaction Functions for the 2018 Tourist Season*

According to Fig. 5.36, the overall satisfaction function based on the 2018 tourist season data has a concave shape, which shows that visitors are not demanding and remain satisfied even if not all their expectations are met.

Table 5.14 MUSA satisfaction results 2020–2022

	2020		2021		2022	
	Level of satisfaction	Weight	Level of satisfaction	Weight	Level of satisfaction	Weight
Total satisfaction	0.916		0.893		0.838	
Tour guides	0.621	7.26	0.604	7.25	0.711	7.68
Signs	0.775	9.22	0.767	8.26	0.813	10.41
Safety	0.853	10.33	0.821	10.02	0.75	10.06
Available time	0.943	27.70	0.947	28.45	0.678	13.35
Landscape	0.895	10.33	0.84	8.41	0.819	12.07
Staff	0.799	8.96	0.82	10.02	0.79	9.09
Local Businesses	0.641	7.42	0.741	10.02	0.829	10.46
Resting areas	0.888	11.11	0.84	10.02	0.887	18.58
Price	0.737	7.67	0.704	7.53	0.733	8.30

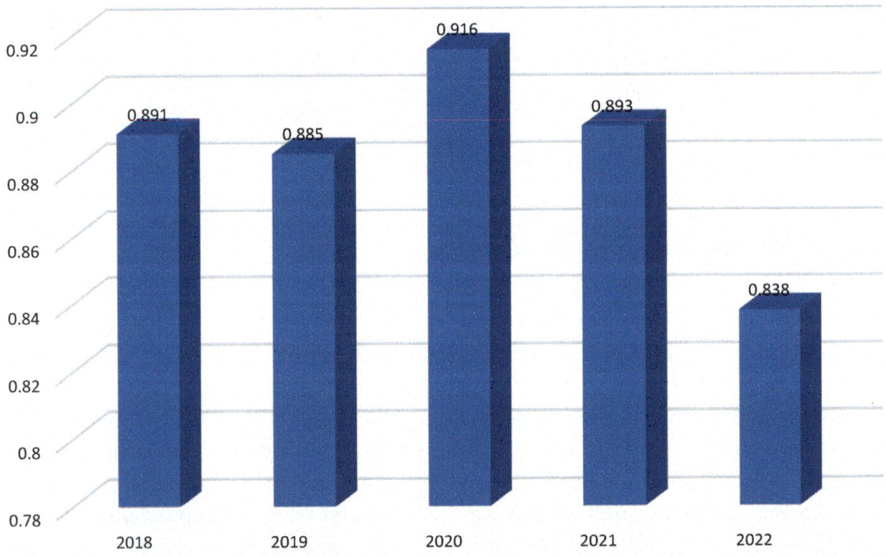

Fig. 5.35 Total satisfaction

In Fig. 5.37 we see that the Satisfaction Function for the criterion “Accompanying tour guides” is almost linear, which suggests that visitors during the 2018 season are neutral with respect to this criterion.

The Satisfaction Function for the criterion “Signage” is also linear, which suggests that visitors during the 2018 season are neutral with respect to this criterion.

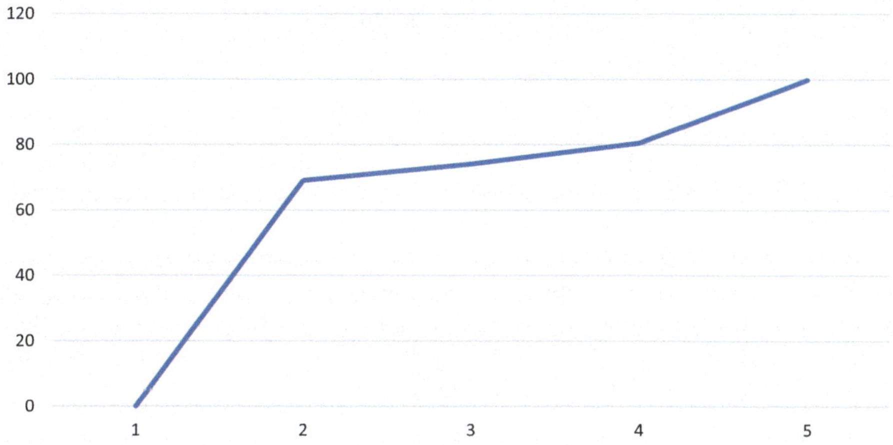

Fig. 5.36 Overall satisfaction function 2018

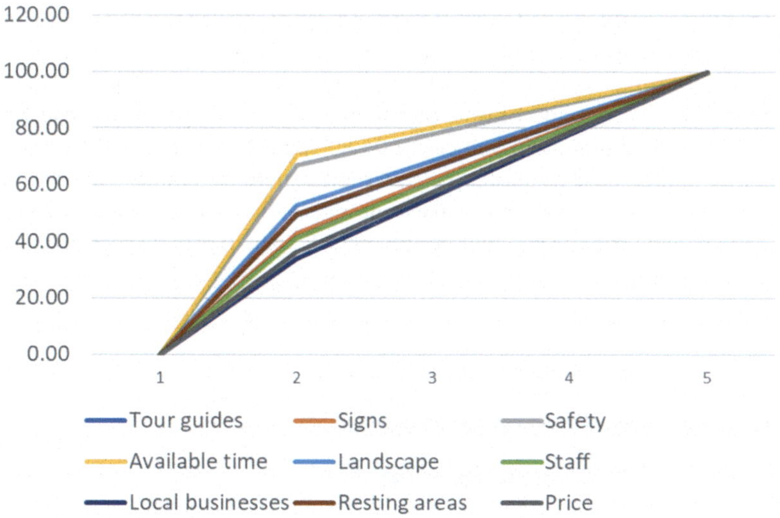

Fig. 5.37 2018 criteria satisfaction functions

The Satisfaction Functions for the criteria "Safety", "Available Time" and "Scenery" are concave, which suggests that visitors during the 2018 season are not demanding with respect to these criteria.

The Satisfaction Functions for the criteria "Staff", "Local Businesses", "Rest Areas" and "Price" are linear, which suggests that visitors during the 2018 season are neutral with respect to these criteria.

5.2.4 *Weighs of Satisfaction Criteria for the 2018 Tourist Season*

The MUSA method weights the role played by each criterion on a scale that sums to 100%. For this weighting, it takes into account the impact of satisfaction with a specific criterion on the overall satisfaction of the visitor and assigns a weight to that criterion.

During 2018 (Fig. 5.38), the available time in the park is the most important criterion for visitors with a weight of 17.92. Safety comes second with a weight of 15.88, the landscape third with a weight of 11.11 and the rest areas located within the park fourth with a weight of 10.43. Local businesses are the criterion with the lowest weight based on 2018 visitor responses.

5.2.5 *Average Satisfaction Indices for the 2018 Tourist Season*

In Fig. 5.39 we see the average satisfaction indices for the criteria based on data for the 2018 tourist season. The highest average satisfaction index concerns the criterion "Available time" with a score of 0.91, indicating that participants were very satisfied with the availability of time during their stay in the park. "Landscape" and "Safety" also scored high, with indices of 0.889 and 0.869 respectively. These areas were highly appreciated by participants. Criteria such as "Guides" and "Signage" had lower satisfaction scores in comparison, with indices of 0.758 and 0.76 respectively. "Staff" and "Rest areas" had indices of 0.771 and 0.805 respectively, indicating a satisfactory performance. "Local businesses" received the lowest satisfaction

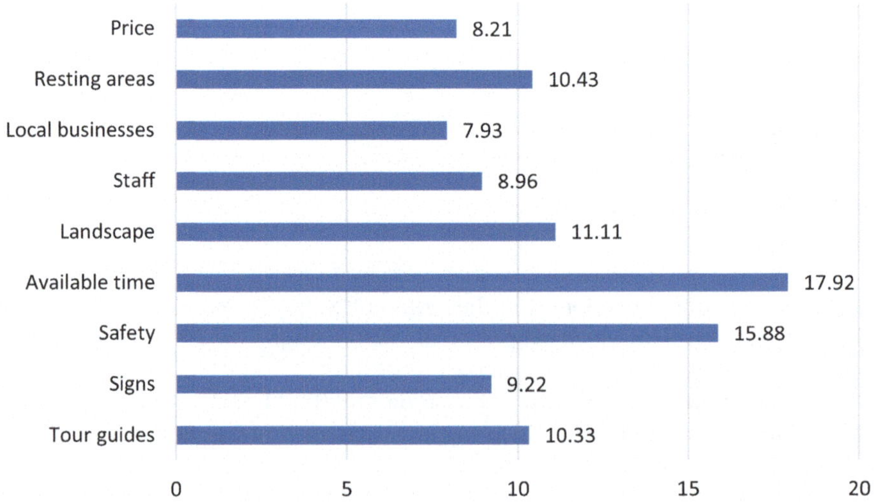

Fig. 5.38 2018 satisfaction criterion weights

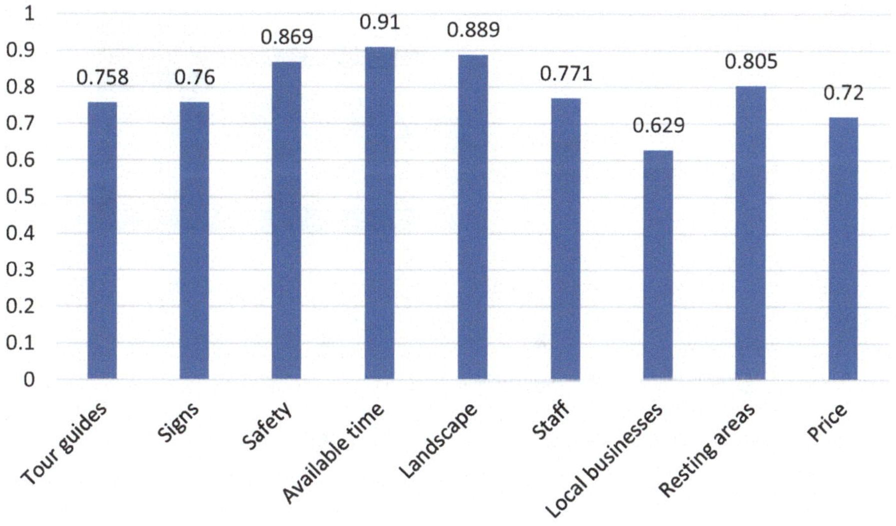

Fig. 5.39 Average satisfaction indices 2018

index with 0.629, indicating that this was the criterion with which the participants were least satisfied. "Price" also had a satisfaction index of 0.72.

5.2.6 Average Demandingness Indices for the 2018 Tourist Season

Figure 5.40 presents the demanding indices for the criteria based on the analysis of data from the 2018 tourist season on a scale from −1 to 1. These indices indicate the measure of effort required to improve each criterion in Samaria National Park. Here are some observations and comments on the graph:

1. Non-Demanding Criteria:
   - Available Time (−0.609): This criterion is the least demanding, with the lowest index, indicating that little or no effort is required for improvement.
   - Safety (−0.559): This criterion also requires little effort for improvement.
   - Landscape (−0.370)
2. Neutral Criteria:
   - Rest areas (−0.329)
   - Staff (0.219)
   - Guides (−0.323)
   - Signage (−0.241)
   - Price (−0.148)
   - Local businesses (−0.117)

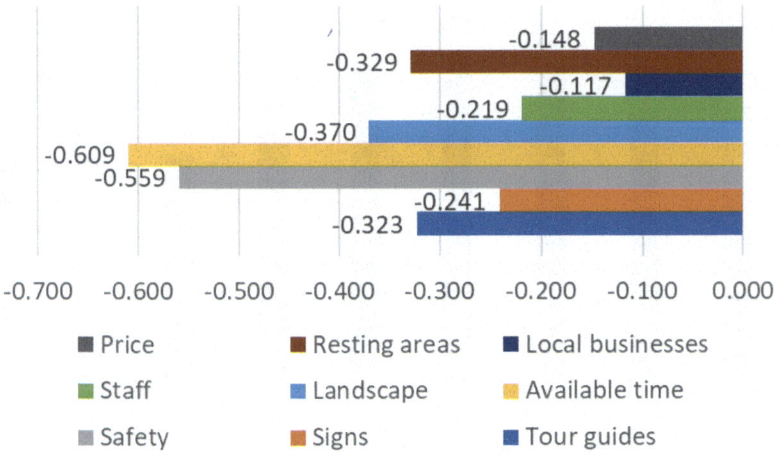

Fig. 5.40 Average demand indicators 2018

3. Demanding Criteria:

• There do not appear to be any demanding criteria.

Therefore, the figure shows that the criteria "Available time", "Safety" and "Landscape" require the least effort for improvement.

5.2.7 Average Efficiency Indices for the 2018 Tourist Season

According to the MUSA results for the 2018 tourist season for the average efficiency indicators as shown in Fig. 5.41, we observe that Local Businesses (0.029) has the highest efficiency indicator, suggesting that any improvements in this attribute will have the greatest impact on customers. Accompanying Guides (0.025) and Price (0.023) also have relatively high indicators, meaning that they are important to customers and improvements in these attributes will be quite effective. Signage (0.022), Safety (0.021), Staff (0.021) and Rest Areas (0.020) have similar effectiveness indices, while Time Available (0.016) and Landscape (0.012) have the lowest effectiveness indices, suggesting that improvements in these will not be as effective compared to the other features.

5.2.8 Satisfaction Functions for the 2019 Tourist Season

According to Fig. 5.42 the overall satisfaction function based on the 2019 tourist season data has a concave shape. This indicates that visitors are not demanding.

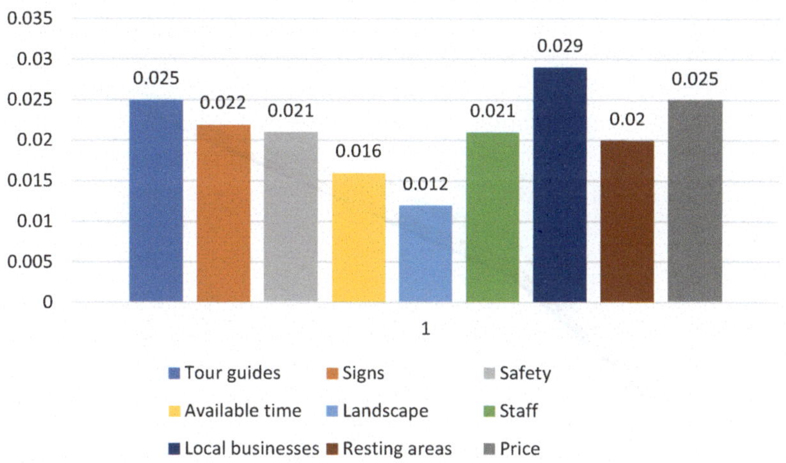

Fig. 5.41 Average effectiveness indices 2018

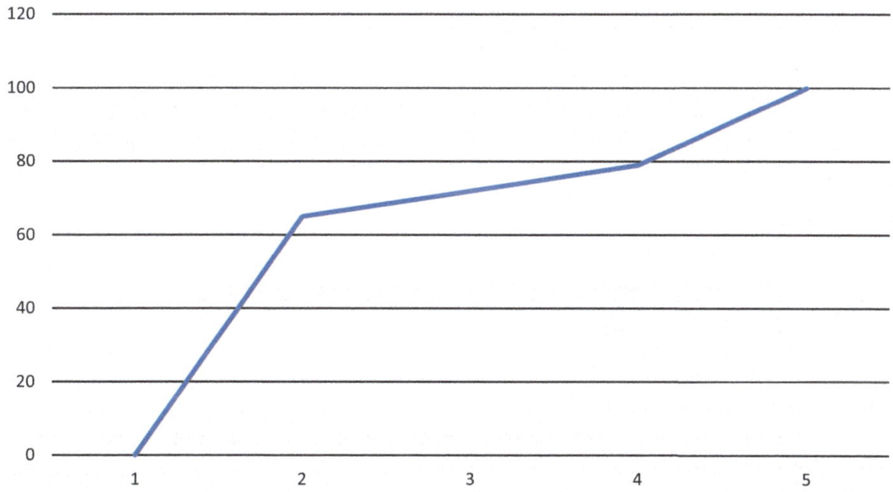

Fig. 5.42 Total satisfaction function 2019

According to Fig. 5.43, the satisfaction function for the criterion "Guides" based on the data of the 2019 tourist season has a linear form. This suggests that visitors are neutral with respect to this criterion.

The satisfaction function for the criterion "Signage" based on the data of the 2019 tourist season also has a linear form, which suggests that visitors are neutral with respect to this criterion as well.

The satisfaction functions for the criteria "Safety", "Available Time", "Landscape" and "Staff" based on the data of the 2019 tourist season have a concave form. This indicates that visitors are not demanding with respect to these criteria.

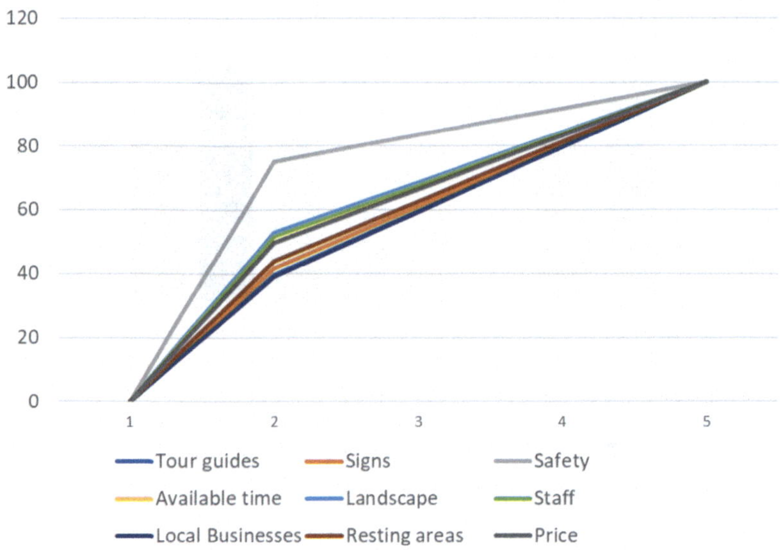

Fig. 5.43 2019 criteria satisfaction functions

The satisfaction functions for the criteria "Local Businesses", "Rest Areas" and "Price" based on the data of the 2019 tourist season have a linear form. This suggests that visitors are neutral towards these criteria.

5.2.9 Weight of Satisfaction Criteria for the 2019 Tourist Season

Analyzing the data from the 2019 tourist season, we observe in Fig. 5.44 that Safety (21.08) has the highest weight, indicating that visitors consider safety to be the most important criterion for their overall satisfaction, while Landscape (11.11) follows. The weights for Time Available (10.89), Staff (10.89) and Price (10.42) indicate that these criteria are moderately important for visitor satisfaction. Signage (9.01) and Rest Areas (9.33) are less important. Guides (8.67) and Local Businesses (8.60) have the lowest weights, indicating that, while they play a role, they are less critical than the other criteria. Overall, Safety and Scenery are the most important criteria for visitors, while Guides and Local Businesses are the least important.

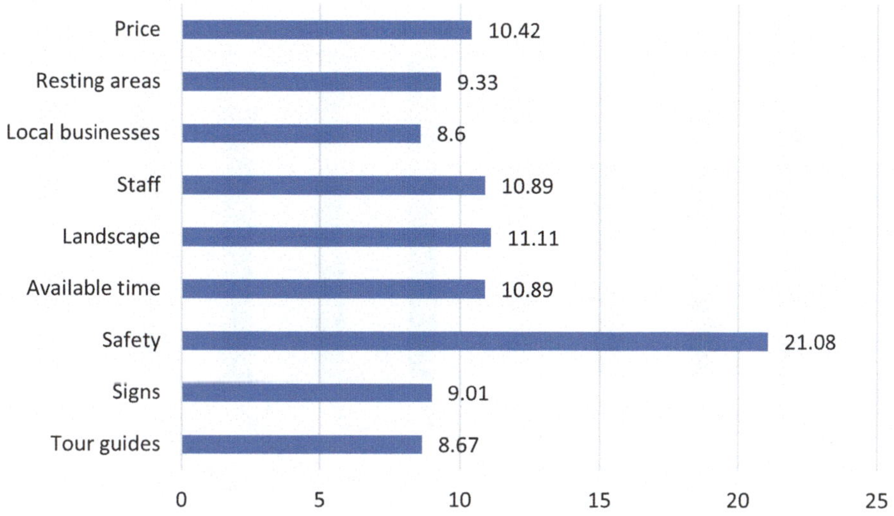

Fig. 5.44 Weights of satisfaction criteria 2019

5.2.10 Average Satisfaction Indices for the 2019 Tourist Season

Figure 5.45 presents the average satisfaction indices for the criteria according to the data from the 2019 tourist season. We observe that Safety (0.901) has the highest satisfaction index, indicating that visitors feel quite safe. Landscape (0.896) and Available time (0.846) also have high satisfaction indices, indicating that visitors are very satisfied with these criteria. Staff (0.817) and Rest areas (0.790) also have significant satisfaction indices. Price (0.785) and Signage (0.746) have slightly lower scores, while Guides (0.721) and Local Businesses (0.660) have the lowest satisfaction scores, indicating that there is room for improvement in these criteria. Overall, Safety and Scenery are the criteria that leave visitors most satisfied, while Local Businesses and Guides need more attention for improvements.

5.2.11 Average Demandingness Indices for the 2019 Tourist Season

In Fig. 5.46 we see the average demandingness indices based on data from the 2019 tourist season:

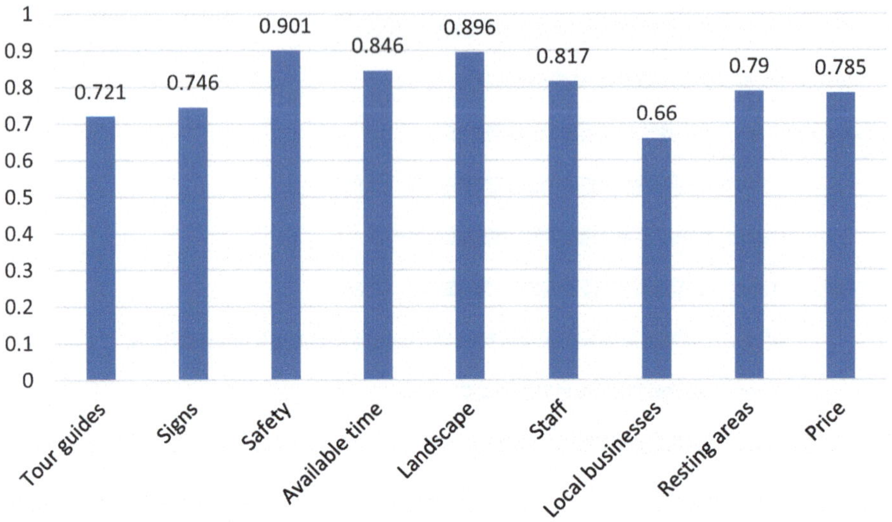

Fig. 5.45 Average criteria satisfaction indices 2019

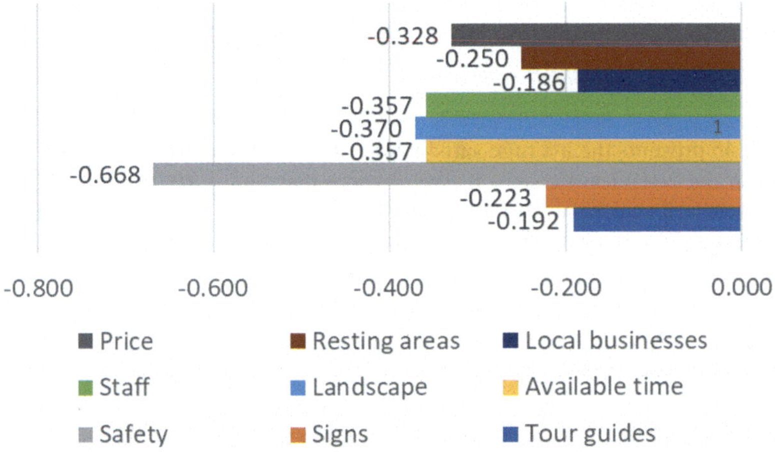

Fig. 5.46 Average demandingness indices 2019

1. The non-demanding criteria are: Safety (−0.668), Available time (−0.357), Landscape (−0.370) and Staff (−0.357). These are the criteria that require the least effort to improve.
2. The neutral criteria are: Price (−0.328), Accompanying guides (−0.192), Signage (−0.223), Local businesses (−0.186) and Rest areas (−0.250).
3. There are no demanding criteria.

5.2.12 *Efficiency Indices for the 2019 Tourist Season*

Analyzing the efficiency indicators as shown in Fig. 5.47, we observe that Local Businesses (0.029) has the highest efficiency indicator, indicating that improvements in local businesses will have the greatest impact on overall visitor satisfaction. Guides (0.024), Signage (0.023) and Price (0.022) are also important criteria, with high efficiency indicators suggesting that improvements in these criteria will be quite efficient. Safety (0.021), Staff (0.020) and Rest Areas (0.020) also show significant efficiency indicators. Available Time (0.017) and Landscape (0.012) have the lowest indicators, indicating that improvements in these criteria will not be as efficient compared to the other criteria.

5.2.13 *Satisfaction Functions for the 2020 Tourist Season*

According to Fig. 5.48 the overall satisfaction function based on the data of the 2020 tourist season has a concave shape. This indicates that visitors are not demanding.

According to Fig. 5.49 the satisfaction functions for the criteria “Guides”, “Signage” and “Safety” based on the data of the 2020 tourist season have a linear shape. This indicates that visitors are neutral towards these criteria.

According to the same figure, the satisfaction functions for the criteria “Available time” and “Resting areas” based on the data of the 2020 tourist season have a concave shape. This indicates that visitors are not demanding regarding these criteria.

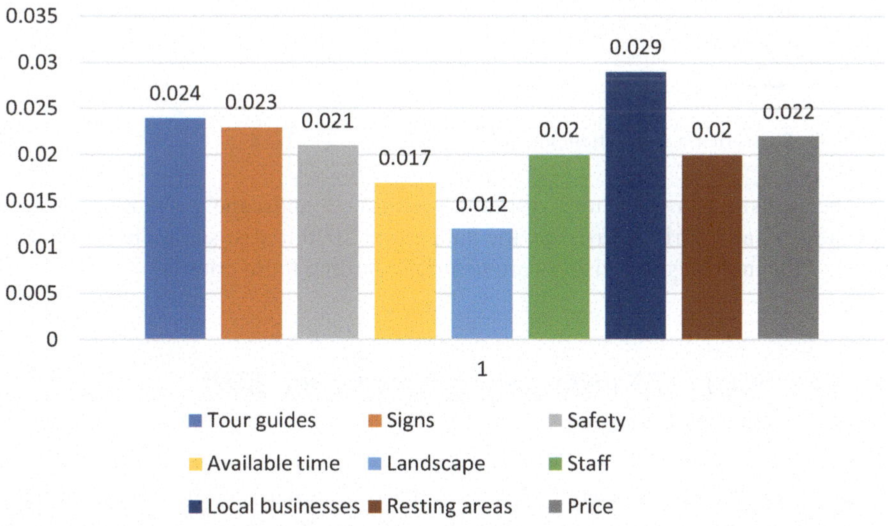

Fig. 5.47 Average efficiency indices 2019

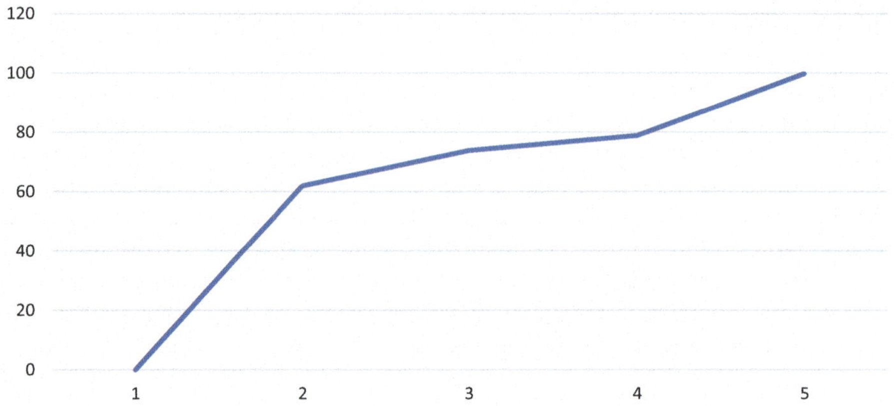

Fig. 5.48 Overall satisfaction function 2020

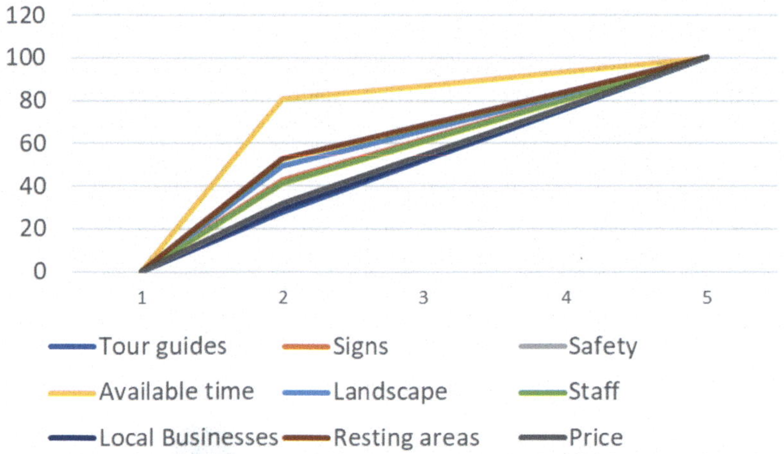

Fig. 5.49 2020 criteria satisfaction functions

The satisfaction functions for the criteria "Landscape", "Staff", "Local Businesses" and "Price" based on the data of the 2020 tourist season have a linear shape. This indicates that visitors are neutral regarding these criteria.

5.2.14 Weight of Satisfaction Criteria for the 2020 Tourist Season

Available time, with a weight of 27.70, was the most important criterion for visitors, as shown in Fig. 5.50. Second in importance were Rest Areas with a weight of 11.11, third Landscape with a weight of 10.33 and Safety with a weight of 10.33. Local businesses and accompanying tour guides were the least important criteria.

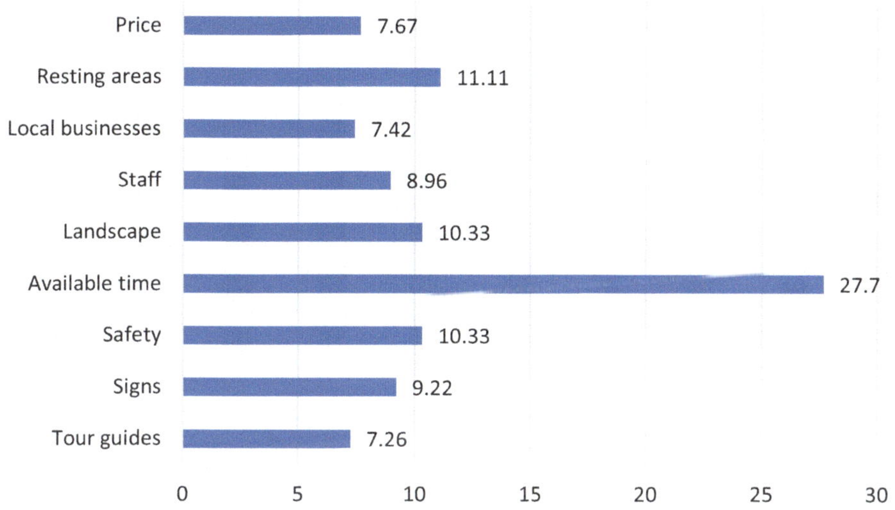

Fig. 5.50 Weights of satisfaction criteria 2020

5.2.15 Average Satisfaction Indices for the 2020 Tourist Season

Analyzing the average satisfaction indices, as shown in Fig. 5.51, we observe that Available time (0.943) has the highest value, indicating that visitors are very satisfied with the time available. Landscape (0.895) and Rest Areas (0.888) also have very high satisfaction indices, indicating that visitors are particularly satisfied with these criteria. Safety (0.853) and Staff (0.799) also show significant satisfaction indicators. Signage (0.775) and Price (0.737) have slightly lower indicators, while Guides (0.621) and Local Businesses (0.641) have the lowest satisfaction indicators, indicating that there is room for improvement in these criteria. Overall, available time, scenery and rest areas are the criteria that visitors seem most satisfied with, while Guides and Local Businesses need more attention for improvements.

5.2.16 Average Demandingness Indicators for the 2020 Tourist Season

In Fig. 5.52 we see the average demandingness indicators for the criteria based on data from the 2020 tourist season.

Analyzing these indicators based on the categorization (from −1 to −0.33 non-demanding, from −0.33 to 0.33 neutral, from 0.33 to 1 demanding), we observe the following:

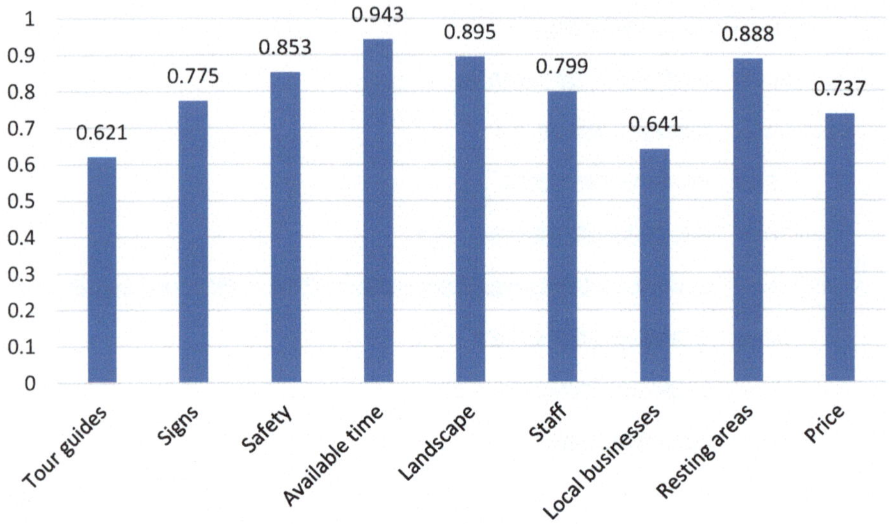

Fig. 5.51 Average satisfaction indices 2020

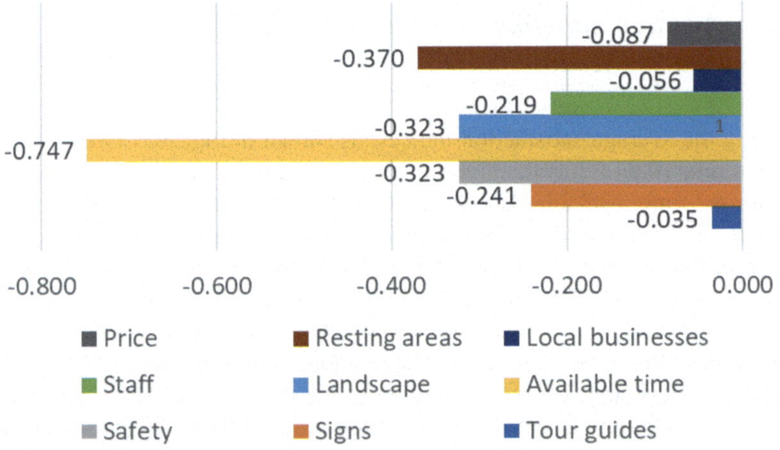

Fig. 5.52 Average demandingness indices 2020

1. Non-demanding criteria:

   - Available time: −0.747
   - Rest areas: −0.370

2. Neutral criteria:

   - Accompanying guides: −0.035
   - Signage: −0.241
   - Safety: −0.323

- Landscape: −0.323
- Staff: −0.219
- Local businesses: −0.056
- Value: −0.087

There are no demanding criteria according to these indicators.

5.2.17 Average Efficiency Indices for the 2020 Tourist Season

In Fig. 5.53 we see the average efficiency indices for the 2020 tourist season. Analyzing these indices, we observe that Accompanying Guides (0.028) and Local Businesses (0.027) have the highest efficiency indicators, indicating that improvements in these characteristics will have the greatest impact on overall visitor satisfaction. Marking (0.021) and Price (0.020) follow, indicating that improvements in these criteria will be efficient. Staff (0.018) and Available Time (0.016) have slightly lower indices. Safety (0.015), Rest Areas (0.012) and Landscape (0.011) have the lowest indices, suggesting that improvements in these criteria will not be as effective compared to the others.

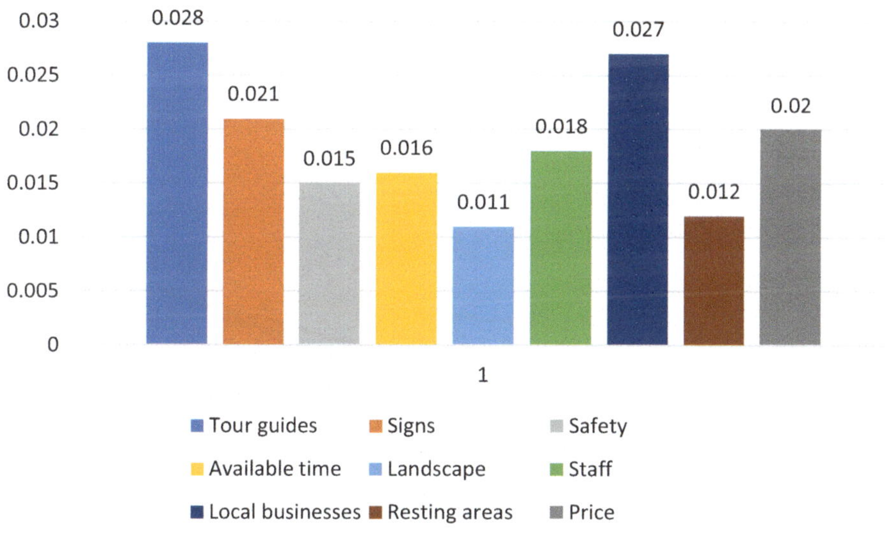

Fig. 5.53 Average efficiency indices 2020

5.2.18 Satisfaction Functions for the 2021 Tourist Season

According to Fig. 5.54, the overall satisfaction function based on the 2021 tourist season data has a linear form, which indicates that visitors are generally neutral.

According to Fig. 5.55, the satisfaction functions for the criteria "Guides", "Signage" and "Safety" based on the data of the 2021 tourist season also have a linear form, which indicates that visitors are generally neutral regarding these criteria.

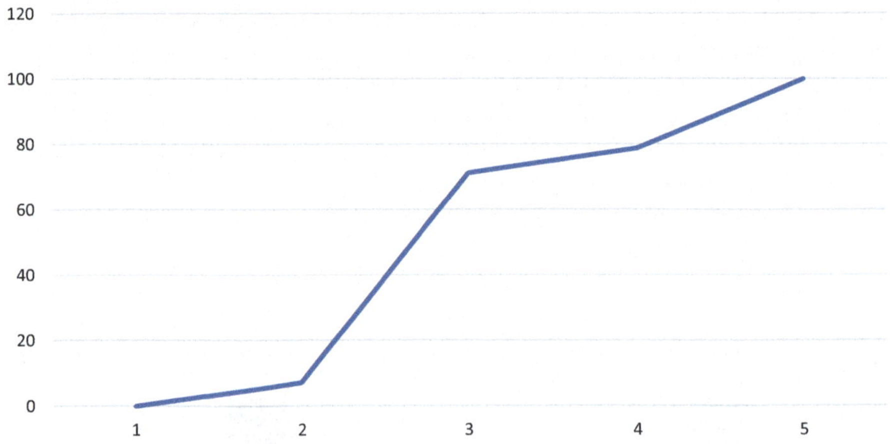

Fig. 5.54 Overall satisfaction function 2021

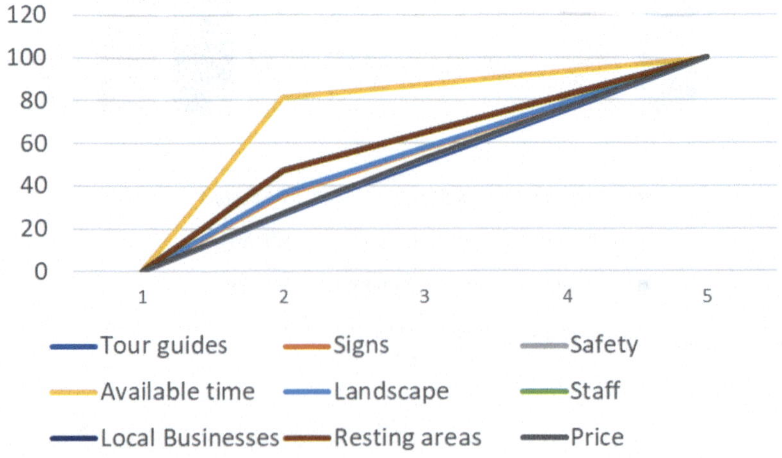

Fig. 5.55 Criteria satisfaction functions 2021

The satisfaction function for the criterion "Available time" based on the data of the 2021 tourist season has a concave form, which indicates that visitors are not demanding regarding the available time in the park.

According to Fig. 5.55, the satisfaction functions for the criteria "Landscape", "Staff", "Local Businesses", "Rest Areas" and "Price" based on data from the 2021 tourist season have a linear form, which indicates that visitors are generally neutral regarding these criteria.

5.2.19 *Weight of Satisfaction Criteria for the 2021 Tourist Season*

In Fig. 5.56 we see that Available time, with a weight of 28.45, was again the most important criterion for visitors. Second were rest areas with a weight of 10.02, Security with a weight of 10.02, park staff with a weight of 10.02 and finally local businesses with a weight of 10.02, which play a much more important role compared to previous years.

5.2.20 *Average Satisfaction Indices for the 2021 Tourist Season*

In Fig. 5.57 we see the average satisfaction indices for the criteria based on data from the 2021 tourist season. Analyzing these indices, we observe that Available Time (0.947) has the highest satisfaction index, indicating that visitors are very

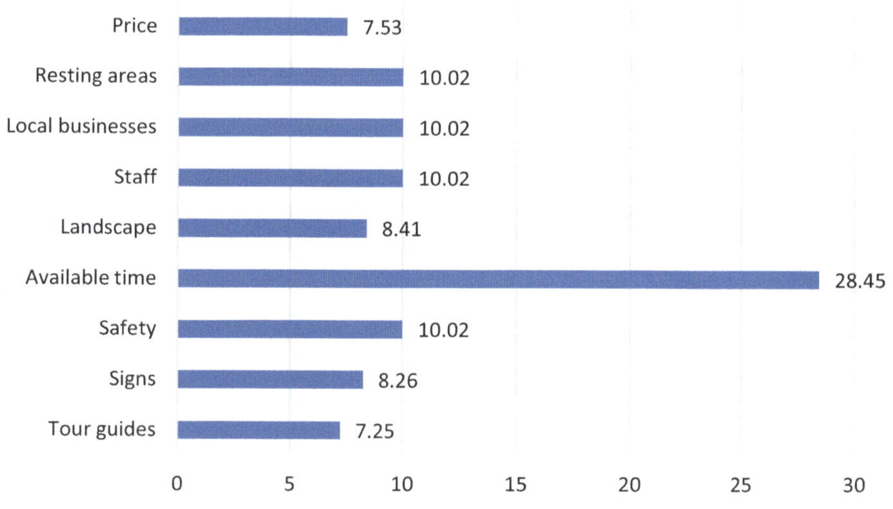

Fig. 5.56 2021 satisfaction criteria weights

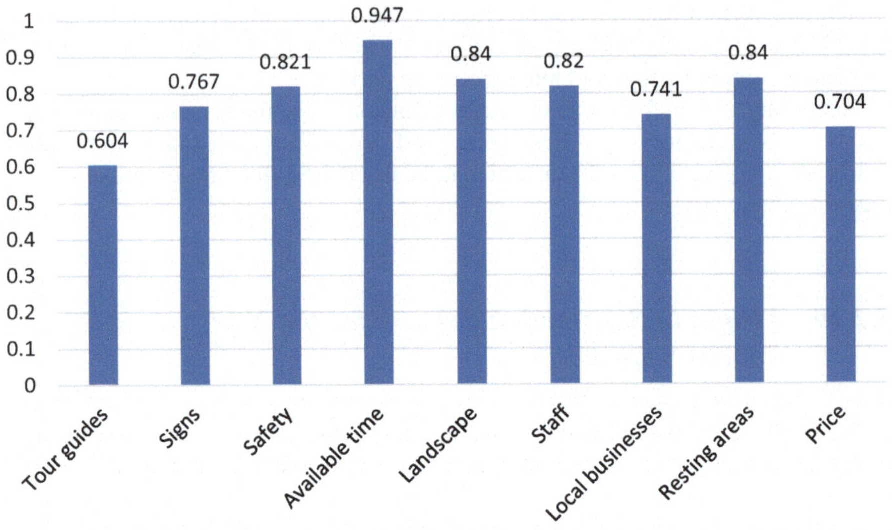

Fig. 5.57 Average criteria satisfaction indices 2021

satisfied with the available time. Landscape (0.840) and Rest Areas (0.840) also have very high satisfaction indices, indicating that visitors are very satisfied with these criteria. Safety (0.821) and Staff (0.820) also show significant satisfaction indices. Marking (0.767) and Price (0.704) have slightly lower scores, while Guides (0.604) and Local Businesses (0.741) have the lowest satisfaction scores, indicating that there is room for improvement in these criteria. Overall, available time, scenery and rest areas are the criteria that visitors are most satisfied with, while Guides and Local Businesses can be improved.

5.2.21 Average Demandingness Indices for the 2021 Tourist Season

In Fig. 5.58 we see the average demandingness indices for the criteria based on 2021 data. Analyzing these indices based on the categorization (from −1 to −0.33 non-demanding, from −0.33 to 0.33 neutral, from 0.33 to 1 demanding), we observe the following:

1. Non-demanding criteria:
    - Available time: −0.751
2. Neutral criteria:
    - Accompanying guides: −0.023
    - Signage: −0.143
    - Safety: −0.294

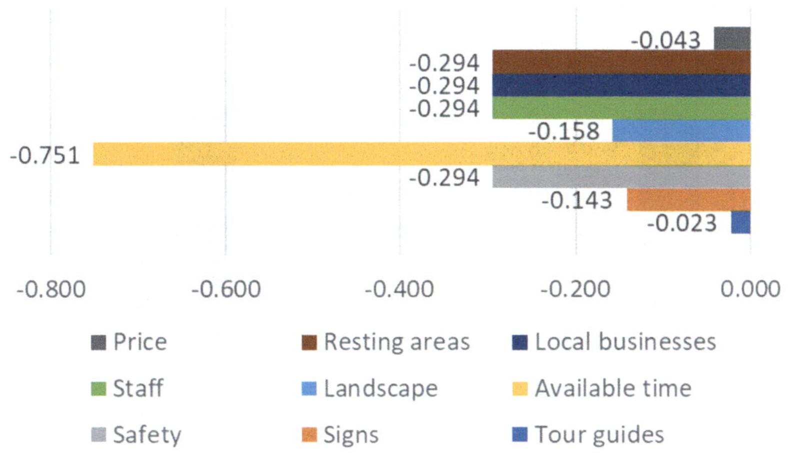

Fig. 5.58 Average demandingness indices 2021

- Landscape: −0.158
- Staff: −0.294
- Local businesses: −0.294
- Rest areas: −0.294
- Price: −0.043

There are no demanding criteria according to these indicators.

5.2.22 Average Efficiency Indices for the 2021 Tourist Season

Figure 5.59 presents the average efficiency indices according to the data from the 2021 tourist season. Analyzing these indices, we observe that Accompanying guides (0.029) and Local businesses (0.026) have the highest efficiency indices, indicating that improvements in these characteristics will have the greatest impact on overall visitor satisfaction. Price (0.022) and Marking (0.019) follow, indicating that improvements in these criteria will be efficient. Staff (0.018) and Safety (0.018) have slightly lower indices. Available Time (0.015) and Rest Areas (0.016) have lower indices, while Landscape (0.013) has the lowest index, suggesting that improvements in these criteria will not be as effective as the others.

5.2.23 Satisfaction Functions for the 2022 Tourist Season

According to Fig. 5.60, the overall satisfaction function based on the 2022 tourist season data has a linear form, which indicates that visitors are generally neutral.

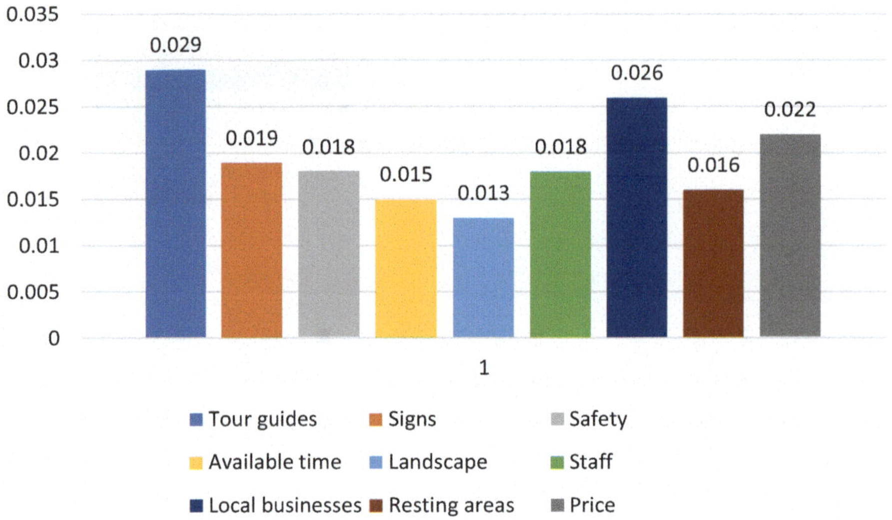

Fig. 5.59 Average efficiency indices 2021

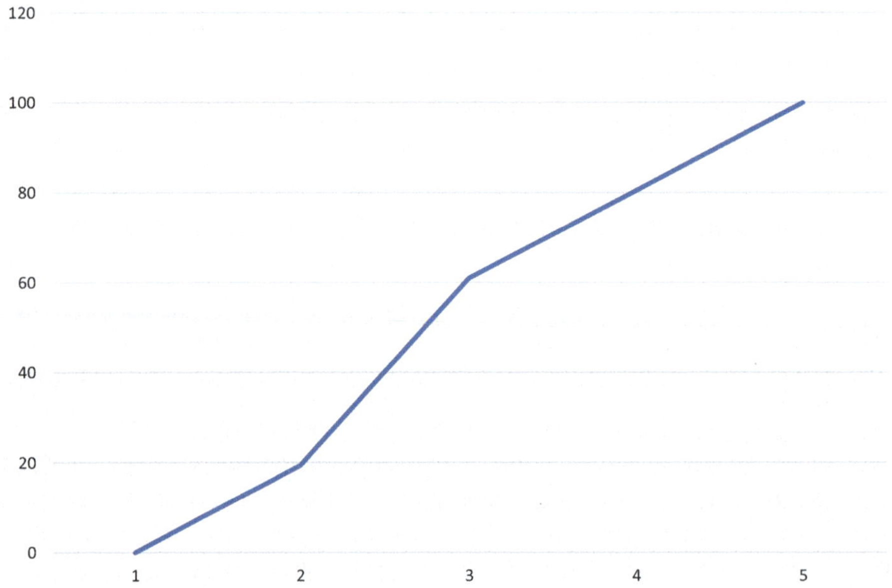

Fig. 5.60 Overall satisfaction function 2022

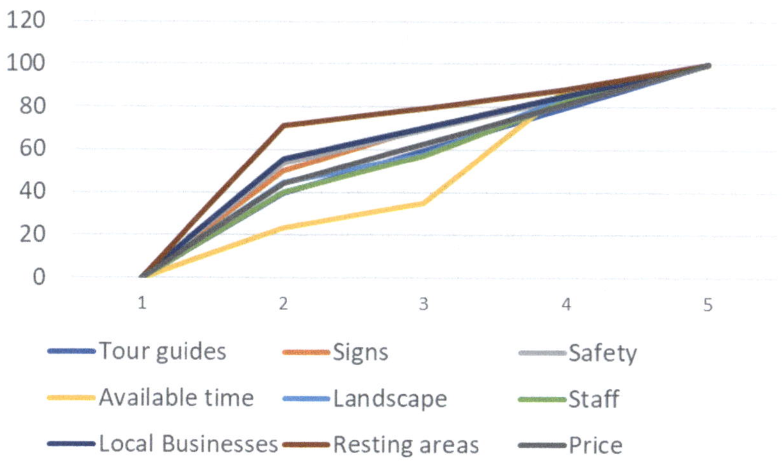

Fig. 5.61 Criteria satisfaction functions 2022

Figure 5.61 shows that the satisfaction function for tour guides based on the 2022 tourist season data also has a linear form, which indicates that visitors are neutral regarding this criterion.

According to the same figure, the satisfaction functions for Signage and Safety based on the 2022 tourist season data have a concave form, which indicates that visitors are not demanding regarding these criteria.

The satisfaction functions for the criteria "Available Time", "Landscape", "Staff" and "Price", based on the 2022 tourist season data have a linear form, which indicates that visitors are neutral regarding these criteria.

The satisfaction functions for Local Businesses and Rest Areas based on the 2022 tourist season data have a concave shape, which indicates that visitors are not demanding in terms of these criteria.

5.2.24 *Weight of Satisfaction Criteria for the 2022 Tourist Season*

In Fig. 5.62 we see that rest areas with a weight of 18.58 were the most important criterion for visitors in 2022. Second was available time with a weight of 13.35, third was Landscape with a weight of 12.07, followed by local businesses with a weight of 10.46, signage with a weight of 10.41 and safety with a weight of 10.06.

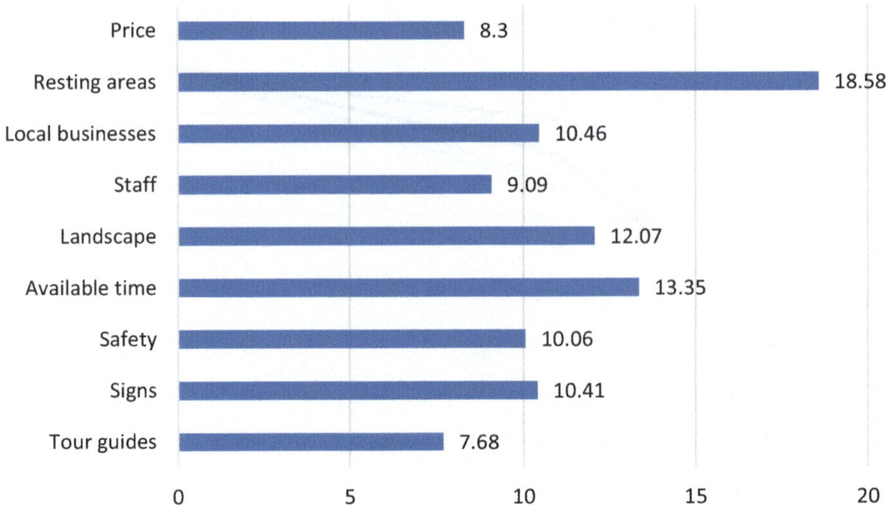

Fig. 5.62 Satisfaction criteria weights 2022

5.2.25 Average Satisfaction Indices for the 2022 Tourist Season

In Fig. 5.63 we see the average satisfaction indices for 2022. Analyzing these indices, we observe that Rest Areas (0.887) has the highest satisfaction index, indicating that visitors are particularly satisfied with this criterion. Local Businesses (0.829) and Landscape (0.819) also have very high satisfaction indices. Signage (0.813) and Staff (0.790) also show significant satisfaction indices. Safety (0.750) and Price (0.733) have slightly lower scores, while Guides (0.711) and Time Available (0.678) have the lowest satisfaction scores, indicating that there is room for improvement in these criteria. Overall, rest areas, local businesses, and scenery are the criteria that achieve the highest average satisfaction scores, while Guides and Time Available appear to need more attention.

5.2.26 Average Demandingness Indices for the 2022 Tourist Season

In Fig. 5.64 we see the average demandingness indices for 2022. Analyzing these indices based on the categorization (from −1 to −0.33 non-demanding, from −0.33 to 0.33 neutral, from 0.33 to 1 demanding), we observe the following:

1. Non-demanding criteria:
   - Rest areas: −0.593
   - Local businesses: −0.407

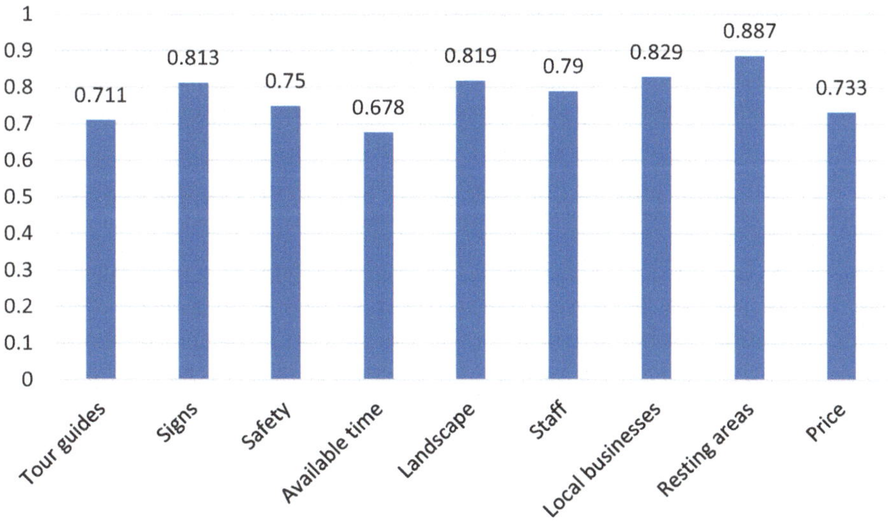

Fig. 5.63 Average criteria satisfaction indices 2022

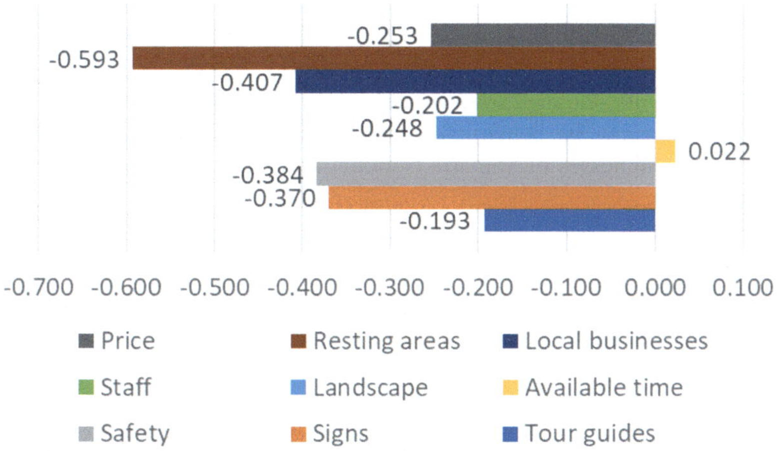

Fig. 5.64 Average demandingness indices 2022

- Safety: −0.384
- Signage: −0.370

2. Neutral criteria:

- Accompanying guides: −0.193
- Landscape: −0.248
- Staff: −0.202
- Price: −0.253
- Time available: 0.022

There are no demanding criteria according to these indices.

Overall, visitors have low demands in relation to rest areas, local businesses, safety and signage, so these criteria can be improved more easily. The remaining criteria are neutral, indicating that visitors have moderate demands in these areas.

5.2.27 *Average Efficiency Indices for the 2022 Tourist Season*

Figure 5.65 presents the average efficiency indices for 2022. Analyzing these indices, we observe that Time Available (0.043) has the highest average efficiency index, indicating that improvements in time available will have the greatest impact on overall visitor satisfaction. Safety (0.025), Guides (0.022), Scenery (0.022), Rest Areas (0.021) and Price (0.022) follow. Signage (0.019) and Staff (0.019) have slightly lower indices, while Local Businesses (0.018) has the lowest efficiency indexx, indicating that improvements in these criteria will not be as effective as the others. Overall, improvements in the available time will have the greatest effectiveness. On the other hand, any improvements in local businesses will have a smaller impact on overall visitor satisfaction.

5.3 Action Diagrams

Based on the results obtained from the analysis of data using the MUSA method, Action Diagrams can be constructed which provide useful information on the strengths and weaknesses of Samaria National Park and on possible opportunities for improvement.

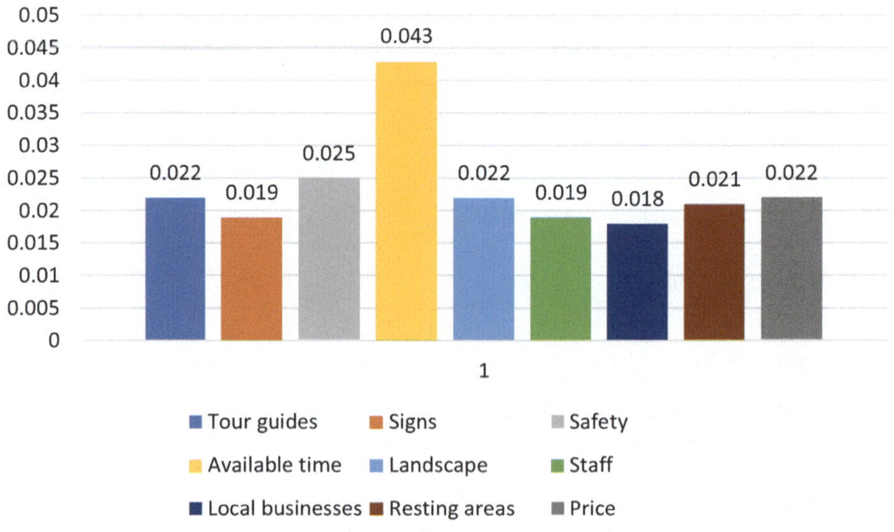

Fig. 5.65 Average effectiveness indices 2022

Action Diagrams are a combination of the weights of the criteria and the corresponding satisfaction indicators, i.e. they show the importance/performance relationship of each criterion. They are similar to the SWOT analysis and can show which elements of satisfaction can be improved and which are the strengths and weaknesses of Samaria National Park. The diagram is divided into 4 squares: the Leverage Opportunity area characterized by high performance and high importance and shows the comparative advantages of Samaria National Park, the Status Quo area characterized by low performance and low importance, the Transfer Resources area characterized by high performance and low importance and shows the possibilities of transferring resources to other elements and the Action Opportunity area characterized by high performance and low importance.

In the 2018 Action diagram (Fig. 5.66), we see:

- Leverage Opportunity area: For this season, the criteria "Available Time" and "Safety" are the comparative advantages of Samaria National Park.
- Status Quo area: For this season, the criteria "Staff", "Signs", "Tour guides", "Price" and "Local businesses" seem to have lower performance but also lower importance for visitors.
- Transfer recourses area: For this season, the criterion "Resting areas" is characterized by high performance but relatively lower importance.
- Action Opportunity area: For this season, there do not seem to be any criteria with high importance and low performance.
- "Landscape" is located on the border between the Leverage Opportunity and Transfer Resources areas

In the 2019 Action Diagram (Fig. 5.67), we see:

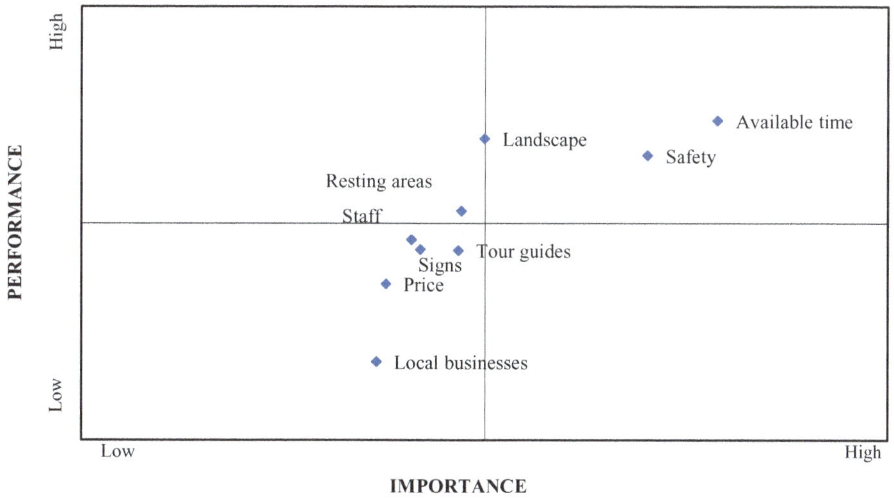

Fig. 5.66 Action diagram 2018

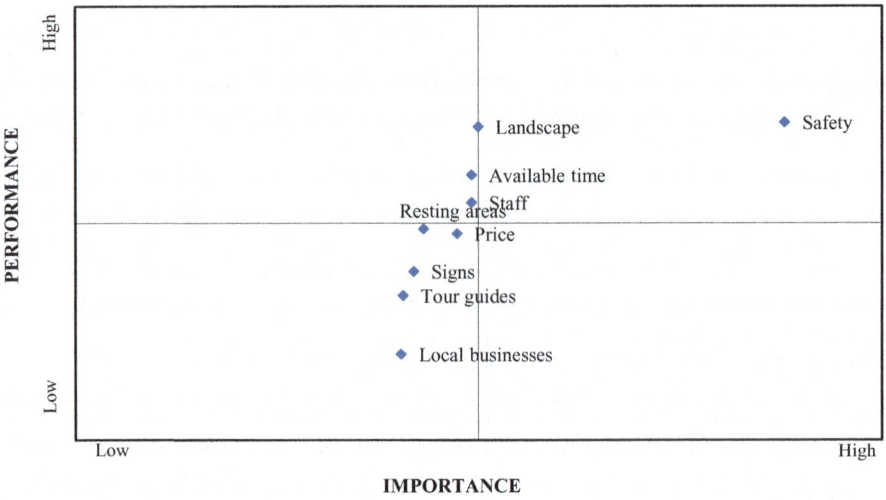

Fig. 5.67 Action diagram 2019

- Leverage Opportunity area: For this season, the criterion "Safety" is the comparative advantage of Samaria National Park.
- Status Quo area: For this season, the criteria "Resting Areas", "Price", "Signs", "Tour guides" and "Local businesses" seem to have lower performance but also lower importance for visitors.
- Transfer recourses area: For this season, the criteria "Available time" and "Staff" are characterized by high performance but relatively lower importance.
- Action Opportunity area: For this season, there do not seem to be any criteria with high importance and low performance.
- "Landscape" is located on the border between the Leverage Opportunity and Transfer Resources areas

In the 2020 Action diagram (Fig. 5.68), we see:

- Leverage Opportunity Area: For this season, the criterion "Available time" is the comparative advantage of Samaria National Park.
- Status Quo Area: For this season, the criteria "Signs", "Price", "Local businesses" and "Tour guides" seem to have lower performance but also lower importance for visitors.
- Transfer recourses Area: For this season, the criteria "Landscape", "Safety" and "Staff" are characterized by high performance but relatively lower importance.
- Action Opportunity Area: For this season, there do not seem to be any criteria with high importance and low performance.
- "Resting areas" is located on the boundary between the Leverage Opportunity and Transfer Resources areas

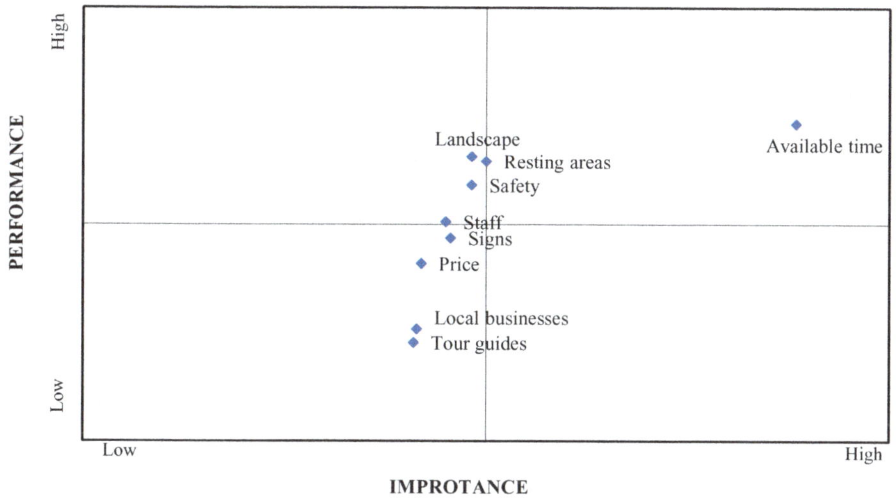

Fig. 5.68 Action diagram 2020

In the 2021 Action Diagram (Fig. 5.69), we see:

- Leverage Opportunity Area: For this season, the criterion "Available time" is the comparative advantage of Samaria National Park.
- Status Quo Area: For this season, the criteria, "Signs", "Tour guides", "Price" and "Local businesses" seem to have lower performance but also lower importance for visitors.
- Transfer recourses Area: For this season, the criteria "Resting areas", "Landscape", "Safety" and "Staff" are characterized by high performance but relatively lower importance.
- Action Opportunity Area: For this season, there do not seem to be any criteria with high importance and low performance.

In the 2022 Action diagram (Fig. 5.70), we see:

- Leverage Opportunity Area: For this season, the criteria "Resting areas" and "Landscape" are the comparative advantages of Samaria National Park.
- Status Quo Area: For this season, the criteria "Safety", "Price" and "Tour guides" seem to have lower performance but also lower importance for visitors.
- Transfer recourses Area: For this season, the criteria "Local businesses", "Signs" and "Staff" are characterized by high performance but relatively lower importance.
- Action Opportunity Area: For this season, the criterion "Available time" is characterized by high importance but relatively lower performance.

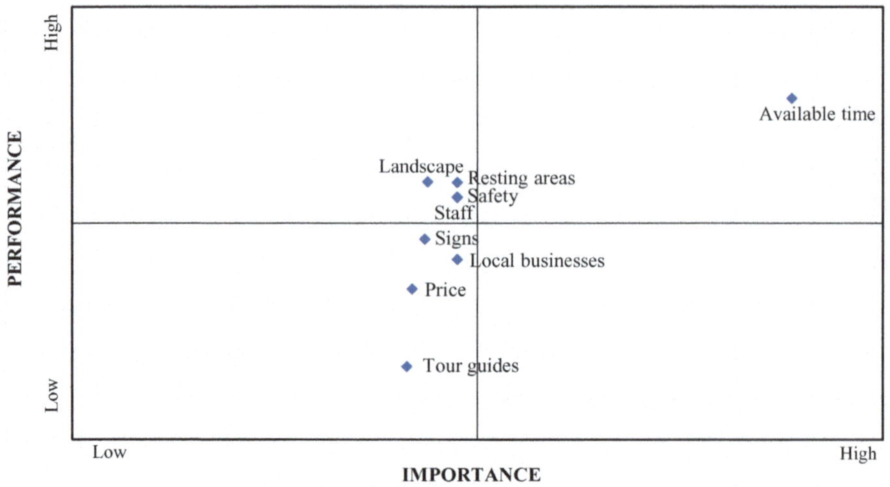

Fig. 5.69 Action diagram 2021

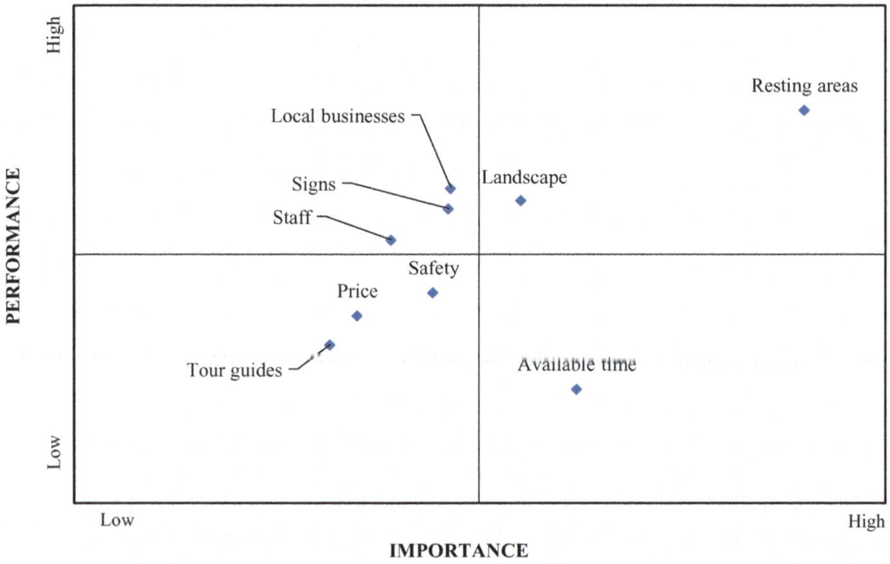

Fig. 5.70 Action diagram 2022

5.4 Improvement Diagram

Improvement diagram are used to categorize opportunities for improvement by considering the demandingness of each criterion for visitors. Each diagram consists of four quadrants based on the combination of demandingness and effectiveness. Higher priority should be given to criteria with high effectiveness and low

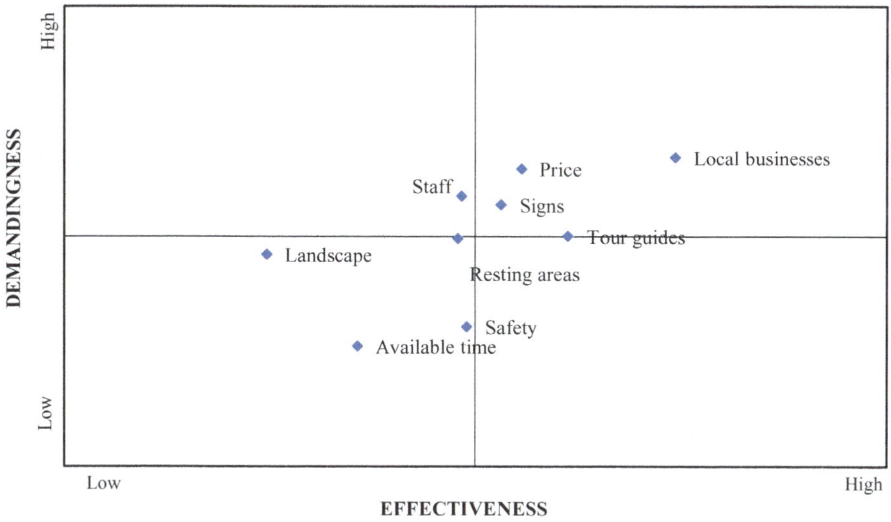

Fig. 5.71 Improvement diagram 2018

demandingness since here there is a high margin for improvement but not much effort is required to achieve it. Conversely, lower importance should be given to criteria with high demandingness and low effectiveness, since they require a lot of effort to achieve improvement (high effort/low effectiveness). In between the two are criteria that are characterized by a high effectiveness index and a high demandingness index and criteria that are characterized by a low effectiveness index and a low demandingness index.

In the 2018 improvement diagram (Fig. 5.71), we see:

- Area characterized by high effectiveness and low demandingness: there do not seem to be any criteria that offer themselves for improvement as a first priority according to the data from the 2018 tourist season.
- Area characterized by high demandingness and high effectiveness: in this area are the criteria "Local businesses", "Price", "Signs" and "Tour guides".
- Area characterized by low demandingness and low effectiveness: in this area are the criteria "Landscape", "Resting areas", "Safety", and "Available time".
- Area characterized by high demandingness and low effectiveness: for this particular season the criterion "Staff" offers less opportunity for improvement.

In the 2019 improvement diagram (Fig. 5.72), we see:

- Area characterized by high effectiveness and low demandingness: "Safety" seems to be on the border of this area.
- Area characterized by high demandingness and high effectiveness: in this area are the criteria "Local businesses", "Price", "Signs" and "Tour guides".
- Area characterized by low demandingness and low effectiveness: in this area are the criteria "Landscape", "Staff" and "Available time".
- Area characterized by high demandingness and low effectiveness: for this particular season the criterion "Resting areas" is less suitable for improvement.

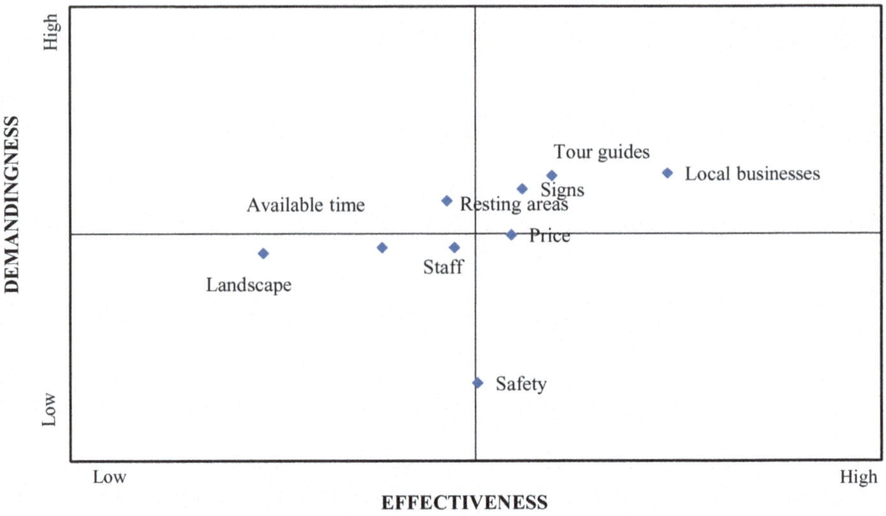

Fig. 5.72 Improvement diagram 2019

In the 2020 improvement diagram (Fig. 5.73), we see:

- Area characterized by high effectiveness and low demandingness: there do not seem to be any criteria that are offered for improvement as a first priority according to the data from the 2020 tourist season.
- Area characterized by high demandingness and high effectiveness: in this area are the criteria "Local businesses", "Price", "Signs" and "Tour guides".
- Area characterized by low demandingness and low effectiveness: in this area are the criteria "Landscape", "Resting areas", "Safety", and "Available time".
- Area characterized by high demandingness and low effectiveness: for this particular season the criterion "Staff" is less suitable for improvement.

In the 2021 improvement diagram (Fig. 5.74), we see:

- Area characterized by high effectiveness and low demandingness: in this area is the criterion "Local businesses" and is offered for improvement as a first priority according to the data from the 2021 tourist season.
- Area characterized by high demandingness and high effectiveness: in this area are the criteria "Price" and "Tour guides".
- Area characterized by low demandingness and low effectiveness: in this area are the criteria "Staff", "Resting areas", "Safety", and "Available time".
- Area characterized by high demandingness and low effectiveness: for this particular season the criteria "Signage" and "Landscape" are less suitable for improvement.

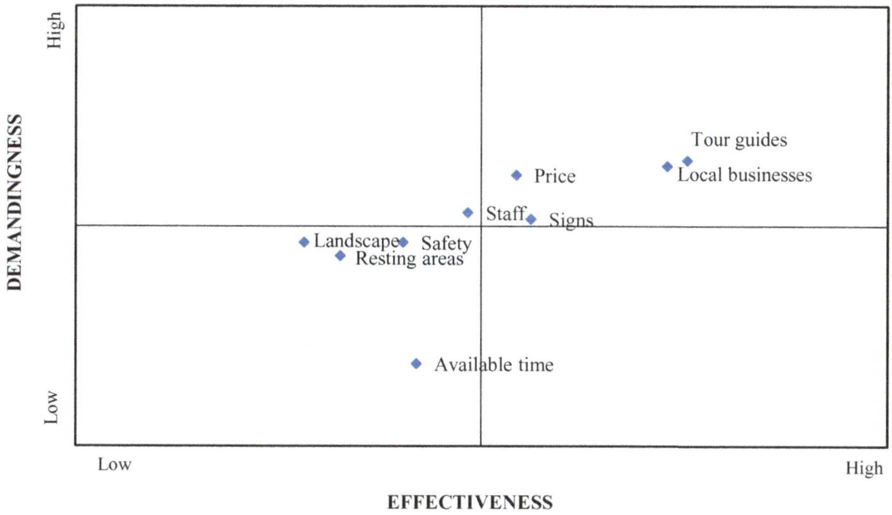

Fig. 5.73 Improvement diagram 2020

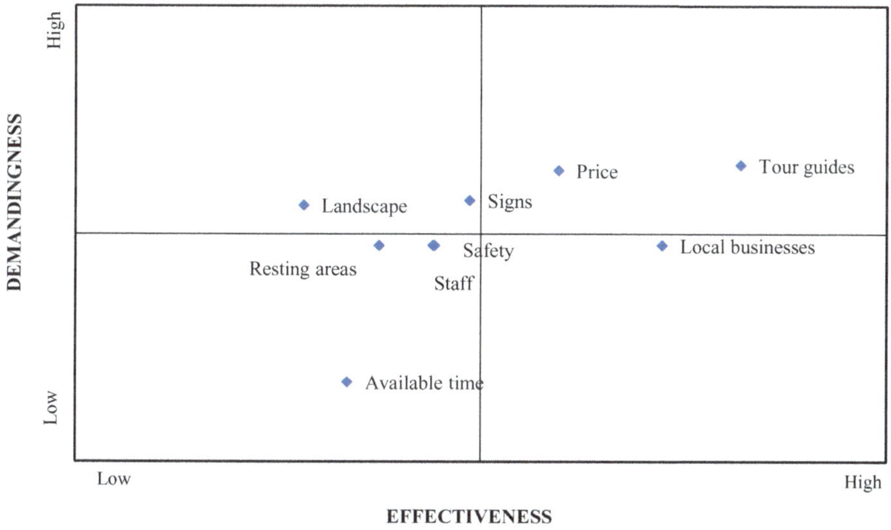

Fig. 5.74 Improvement diagram 2021

In the 2022 improvement diagram (Fig. 5.75), we see:

- Area characterized by high effectiveness and low demandingness: the criterion "Available time" is the first priority for improvement according to the data from the 2022 tourist season.
- Area characterized by high demandingness and high effectiveness: the criterion "Safety" is located in this area.

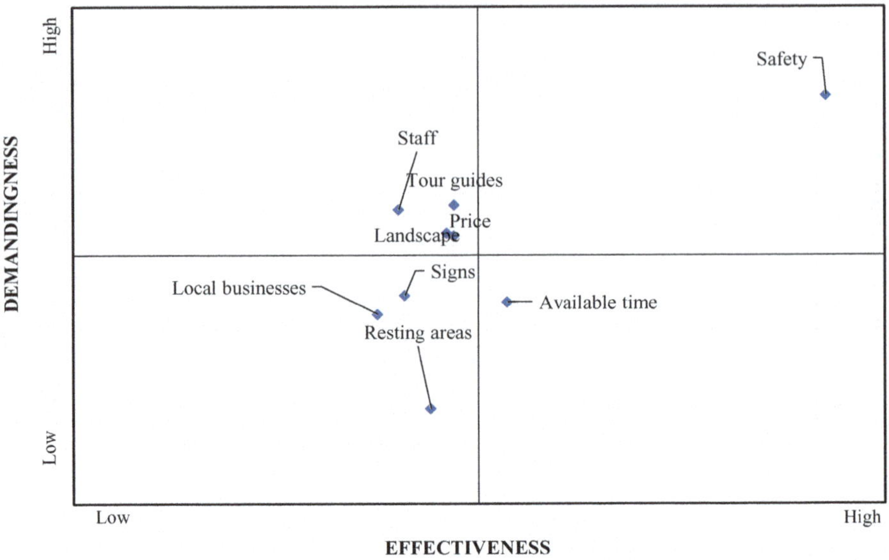

Fig. 5.75 Improvement diagram 2022

- Area characterized by low demandingness and low effectiveness: the criteria "Local businesses", "Resting areas" and "Signs" are located in this area.
- Area characterized by high demandingness and low effectiveness: for this season the criteria "Staff", "Tour guides", "Landscape" and "Price" are the least suitable for improvement.

5.5 General Comments on the Results

Tourism in national parks and protected areas is an important component of the global tourism product, attracting millions of visitors each year. These areas offer unique opportunities for recreation, education and connection with nature, while also contributing significantly to the economic development of local communities. Revenues from tourism play a vital role in the effort to protect nature and maintain biodiversity, financing park management, wildlife protection and ecosystem restoration. In addition, visitors are informed about their importance through educational programs and activities, cultivating a culture of respect and responsibility towards nature.

Tourism activities in or around national parks and protected areas include a wide range of options that meet the preferences and needs of different groups of visitors. Ecotourism, for example, focuses on visits to natural areas that are conducted in a responsible manner with the aim of conserving the environment and enhancing the well-being of the people living in those areas. Tours to areas that are wildlife sanctuaries give visitors the opportunity to observe and photograph wildlife in their natural environment, raising awareness of the need to conserve these species. Hiking and mountaineering on trails that offer visitors the opportunity to enjoy impressive landscapes and have opportunities for physical exercise are also popular activities. Camping in specially designed areas allows visitors to experience nature up close, while marine activities, such as diving and snorkeling, offer unique experiences in coastal and marine protected areas. Considering the fact that tourism in the area around Samaria National Park is usually focused on crossing the main gorge trail during the months it is open, it becomes obvious that there is room to develop quite a few other related activities even during the winter period.

However, tourism in protected areas and national parks poses several challenges. Excessive visitor numbers can damage ecosystems, reduce the quality of the visitor experience due to overcrowding, and lead to the degradation of natural resources. Soil erosion, pollution, and poorly planned infrastructure construction can also have negative impacts on the environment. A proper visitor management plan is critical to maintaining the sustainability of national parks. This may include regulating visitor numbers, establishing use zones, implementing policies that promote conservation and responsible use of resources, and improving features that affect visitor satisfaction. The involvement of local communities in the management of national parks is also crucial. Local communities can benefit from the economic opportunities that tourism creates, while at the same time actively participating in the conservation of natural resources. Education and training programs for local communities can enhance their participation and improve the services and products provided.

Technology and innovation can also play an important role in tourism in national parks. The use of digital platforms through which visitors can book tickets, receive information and be guided, improves their experience and facilitates the management of parks. Tracking systems using GPS and other technologies can help manage visitor and wildlife movements, allowing for informed management decisions. The development of digital educational materials and the implementation of webinars contribute to raising public awareness so that all people understand the importance of preserving national parks.

The use of visitor experience evaluation tools that will provide the necessary information for improvement actions is also very important. The satisfaction of visitors to national parks is a critical indicator of the success and effectiveness of the management of these areas. Park managers and researchers often measure visitor satisfaction to assess the quality of services and infrastructure, the visitor experience, and the sustainability of their activities. Visitor satisfaction is influenced by many factors. Infrastructure and facilities play an important role, as well-maintained trails, campsites, restrooms, and visitor centers are critical to visitor comfort. Accessibility for people with disabilities and facilities for children and families also enhance the experience. The quality of services is equally important. The friendliness and knowledge of staff, the availability of guides, and the provision of information positively affect satisfaction. Well-organized tours and educational programs enhance the visitor experience. The natural environment and landscape are also key elements. Preserving natural beauty and minimizing human intervention contribute to a positive experience.

Several methods are used to measure visitor satisfaction. Surveys and questionnaires ask guests to rate their experience, while face-to-face interviews and focus group discussions provide qualitative information about guest perceptions and experiences. Social media-based analytics provide direct and unsolicited guest opinions.

Visitor satisfaction is crucial to the long-term success of national parks and protected areas. By continuously monitoring and improving services, infrastructure and educational programs, national parks can provide unforgettable experiences for their visitors, while at the same time protecting and conserving natural resources for future generations, and contributing to this goal was the main purpose of this research.

During the years in which the research was conducted, the world was hit by the COVID-19 pandemic, which had a significant impact on tourism in national parks around the world. Many national parks closed completely for periods or limited the number of visitors to prevent the spread of the virus. International travel restrictions and lockdowns have dramatically reduced the number of tourists from abroad, leading to a decline in visitors. The decline in visitor numbers has had an economic impact, as revenues from entrance fees, permits, and services have declined, affecting funding for park maintenance and management. In addition, many national park employees, as well as employees of local businesses that serve tourists, have faced reduced hours or layoffs. Despite the negative economic impact, there have also been positive environmental impacts. The decline in visitor numbers has led to less pressure on natural ecosystems, giving the environment time to recover from overuse and pollution. Some parks have seen an increase in wildlife and improved ecosystems due to reduced human activity. The pandemic has also changed visitor behavior. With international travel limited, more people have chosen to visit national parks in their own countries. People, seeking safe outdoor activities, turned increasingly to national parks for hiking, camping, and other outdoor activities.

National parks have been forced to adapt to new challenges, implementing strict health protocols. In some places, it was necessary for visitors to wear masks, maintain social distancing, and disinfect areas. Some parks developed digital tours and educational materials to continue to educate and inspire the public despite the restrictions. The COVID-19 pandemic has forced national parks to adapt quickly to new challenges, but it has also provided opportunities to recover and rethink tourism management practices.

The study of visitor satisfaction in Samaria National Park is very important for understanding the effectiveness of existing management practices and developing strategies to improve the visitor experience. The analysis of the factors that influence satisfaction provides valuable information about the needs and expectations of visitors, while the continuous monitoring and improvement of infrastructure, services and management practices.

5.6 Comments about the Visitors

The findings of the survey specifically revealed the following about visitors:

Profile

- Gender Distribution: Over the 5-year period, the gender distribution of visitors remained balanced, with small fluctuations each year. No significant differences were observed in the number of men and women.
- Education Level: Visitors of all levels of education were recorded, with a significant number of visitors holding university degrees or higher education (master's or doctorate).
- Visitor Age: The age groups of visitors ranged widely, with a notable increase in visitors aged 36–50 in 2022. There was also a significant increase in visitors over 65, indicating the park's appeal to various age groups.
- Annual Income: The income levels of visitors showed diversity, with fluctuations in various income categories over the years. There was a significant increase in visitors with lower income levels (up to €10,000) in 2020, possibly due to the economic impact of the pandemic. It may also have played a major role that during this particular tourist season, domestic tourism moved more and the percentage of Greek visitors was significantly higher.
- Nationality: The majority of visitors came from abroad, except in 2020, where Greek visitors were significantly more due to travel restrictions and the fear caused by the COVID-19 pandemic.

Characteristics of the visit

- Sources of information: Most visitors learned about the park through the internet and recommendations from friends and relatives. A small percentage of visitors arrived at the park without prior knowledge before visiting Crete.
- Means of Transport: Tourist agency buses were the most popular means of transport in the park, followed by local public transport buses and private cars. Due to the pandemic, there was an increase in the preference for more personal means of transport, such as cars, especially in 2021.
- Repeat visit: The majority of visitors state that they are visiting Samaria National Park for the first time, although there was an increase in repeat visitors during the pandemic years due to the increase in domestic tourism.
- Impact of the Park on Travel Decision: The presence of the park significantly influenced the travel decisions of a significant number of visitors.
- Willingness to Pay: A large percentage of visitors are willing to pay up to €5 more. The willingness to pay a higher entrance fee was generally stable, indicating the perceived value of the park.

Figure 5.76 presents the average rating of how likely visitors to Samaria National Park are to recommend the visit to friends and relatives, for the years 2018–2022, which remains high throughout the 5-year period, with all values being above 4.4. This indicates a stable satisfaction of visitors with the services and experiences offered by the park.

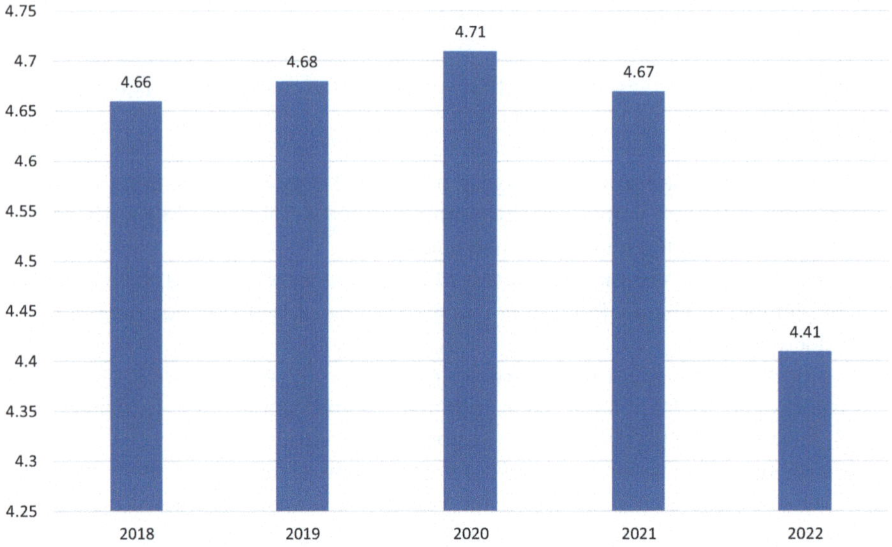

Fig. 5.76 Encouraging friends and relatives to visit Samaria National Park

2020 was the year with the highest rating (4.71), which shows that visitors that year were very satisfied with their experience. The small fluctuations in the ratings in the following years do not negate the fact that it remains at high levels, indicating that the park continues to offer unforgettable experiences.

Maintaining these high ratings is a significant achievement and shows that visitors appreciate the experience offered by Samaria National Park. This positive feedback can be used to continuously promote and improve the park, offering even better services in the future.

5.7 Comments About the Results of the Application of MUSA Method

The application of the MUSA method showed:

Overall Satisfaction

- Overall Satisfaction Level: It remained relatively high for all 5 years of the survey, with small fluctuations. The highest level was recorded in 2020 (0.916), while the lowest in 2022 (0.838).
- Satisfaction Function: The Satisfaction Function had a concave shape during the years 2018–2020 indicating non-demanding visitors who are generally characterized by high levels of satisfaction even when their expectations are not met to a large extent. During the years 2021 and 2022 the Satisfaction Function had a linear shape indicating visitors who the more their expectations are met, the more satisfied they are.

- Conclusion: Overall visitor satisfaction was consistently high, with a slight drop in 2022, likely due to overcrowding due to the very high number of visitors recorded during the season combined with other factors such as visitor attitudes and perceptions after the pandemic. Visitors generally appear to be undemanding but there is a trend of reversal in this

Tour Guides

- Satisfaction Level: Satisfaction decreased from 0.758 in 2018 to 0.604 in 2021, but recovered to 0.711 in 2022.
- Weight: The weight of the criterion ranged between 7.248 and 10.333.
- Demandingness Indices: Visitors were generally neutral towards the tour guides.
- Effectiveness Indices: Improvements in the tour guides would have a moderate impact in absolute terms on increasing overall satisfaction.
- Action diagram: This criterion is in the Status Quo region for all years of the survey, indicating low relative performance and relative importance.
- Improvement diagram: From 2018 to 2021 it is in the area of high demandingness and high effectiveness, ranking it in the second priority of factors that can be improved. In 2022 it moves to the area of high demandingness and low effectiveness, ranking it in the third priority of factors that can be improved.
- Conclusion: Satisfaction with the tour guides was relatively lower compared to other criteria, indicating that there is room for improvement in the quality of their services, but the weight of the criterion was comparatively low for each year, indicating that they do not significantly affect the overall level of visitor satisfaction, while any improvement would not significantly affect overall satisfaction due to the low efficiency indicators and the neutrality of visitors in terms of demand.

Signs

- Satisfaction Level: Satisfaction remained stable and relatively high, with small fluctuations, reaching 0.813 in 2022.
- Weight: The weight was relatively low, reaching its highest point in 2022 (10,412).
- Demandingness Indices: Visitors were generally neutral or undemanding regarding signage.
- Effectiveness Indices: Improvements in signage would have a moderate impact in absolute terms on increasing overall satisfaction.
- Action diagram: For the years 2018–2021 it is in the Status Quo area, showing low performance and importance. In 2022 it moves to the Transfer Resources area, showing higher performance and lower importance.
- Improvement diagram: From 2018 to 2020 it is in the area of high demandingness and high effectiveness, ranking it in the second priority of factors that can be improved. In 2021 it moves to the area of high demandingness and low effectiveness while in 2022 it is in the area of low demandingness and low effectiveness.
- Conclusion: The signs along the path were considered satisfactory by visitors and satisfaction was relatively high, but based on the weight of the criterion

which was comparatively low, we conclude that as a criterion it did not significantly affect the overall satisfaction of visitors in any of the tourist seasons during which the research was carried out. Any improvement would not significantly affect the overall satisfaction due to both the low effectiveness indices and the neutrality of visitors in terms of demand.

Safety

- Satisfaction Level: High satisfaction throughout the survey period, peaking in 2019 (0.901) and declining in 2022 (0.750).
- Weight: The weight of the criterion was high in 2018 and 2019, but decreased in subsequent years.
- Demandingness indices: Visitors were generally undemanding or neutral towards safety.
- Effectiveness indices: Improvements in safety would have a moderate impact in absolute terms on increasing overall satisfaction.
- Action diagram: For 2018 and 2019 it is in the Leverage Opportunity area indicating high performance and importance. In 2020 and 2021 it moves to the Transfer recourses area indicating high performance but lower importance. In 2022 it is in the Status Quo area indicating lower performance and importance for visitors this year.
- Improvement diagram: 2018 is in the low demandingness and low effectiveness area. 2019 is on the border of the low demandingness and high effectiveness area. 2020 and 2021 are again in the low demandingness and low effectiveness area while 2022 moves to the high demandingness and high effectiveness area.
- Conclusion: Safety remained at satisfactory levels for visitors, although satisfaction decreased in 2022. During the years 2018 and 2019, the weight of the criterion indicates that it was a criterion that significantly influenced the overall satisfaction of visitors. On the other hand, during the years 2020–2022, the weight was significantly lower, indicating that the overall satisfaction of visitors for these seasons was more influenced by other criteria.

Available Time

- Satisfaction Level: Very high level of satisfaction, peaking in 2021 (0.947), but a significant decrease in 2022 (0.678).
- Weight: The weight was high, especially in the years 2018, 2020, 2021 and 2022.
- Demandingness indices: Visitors were undemanding towards available time, but in 2022 there was a change to the positive side.
- Effectiveness indices: Improvements in available time would have a greater impact on increasing overall satisfaction in 2022.
- Action diagram: In 2018 is in the Leverage Opportunity area indicating high performance and importance. In 2019 moves into the Transfer recourses area indicating high performance but lower importance. In 2020 and 2021 it is again in the Leverage Opportunity area while in 2022 it moves to the Action Opportunity area indicating low performance and high importance.
- Improvement diagram: From 2018 to 2021 it is consistently in the low demandingness and low effectiveness area. In 2022 it moves to the low demandingness and high effectiveness area.

- Conclusion: Satisfaction with available time was quite high, but decreased significantly in 2022 probably due to overcrowding due to the significantly higher number of visitors, which led visitors to feel that they did not have enough time to enjoy the space as they would like. The high weight attributed to this criterion shows that available time in the park is a fixed factor that significantly affects overall visitor satisfaction. Its highest value was recorded during the years 2020 and 2021, which were the ones most affected by the pandemic, highlighting the importance that visitors gave to the opportunity that a visit to the Samaria National Park provided to them to be in an open space in nature. At the same time, the fact that the maximum level of satisfaction with this criterion was observed during the years in which the number of visitors was the lowest, may be another element indicating that there is a relationship between the evaluation of the available time and therefore its impact on the experience gained by visitors and the number of visitors.

Landscape

- Satisfaction Level: Consistently high satisfaction, with small fluctuations from 0.889 in 2018 to 0.819 in 2022.
- Weight: The weight remained relatively low but increased in 2022 when it became the third most important criterion.
- Demandingness indices: Visitors were non-demanding or neutral towards the landscape, with small fluctuations over the years.
- Effectiveness indices: Improvements to the landscape would have a moderate impact in absolute terms on increasing overall satisfaction.
- Action diagram: In 2018 and 2019, Landscape is on the border between the Leverage Opportunity and Transfer Resources areas, indicating high performance and importance. In 2020 and 2021, it moves into the Transfer Resources area, indicating high performance but lower importance. In 2022, it moves back to the Leverage Opportunity area.
- Improvement diagram: From 2018 to 2020, it is in the low demandingness and low effectiveness area. In 2021 and 2022, it moves to the high demandingness and low effectiveness area.
- Conclusion: Satisfaction from the landscape remained consistently high, indicating the undeniable natural beauty of the park and the fulfillment of visitors' expectations regarding it. The drop in 2022 may be related to the large number of visitors and the overcrowding that that may have been created, which may not have allowed visitors to enjoy it as they would have liked.

Staff

- Satisfaction Level: Satisfaction increased slightly in 2019 (0.817) and remained relatively stable in the following years.
- Weight: The weight ranged from 8960 in 2018 to 10,889 in 2019 with values between these two in the remaining years.
- Demandingness indices: Guests were generally not demanding or neutral towards the staff.

- Effectiveness indices: Improvements in staff would have a moderate impact in absolute terms on increasing overall satisfaction.
- Action diagram: In 2018 is in the Status Quo area indicating low importance and low performance. During the following years from 2019 to 2022 it moves to the Transfer Resources area indicating low importance but higher performance.
- Improvement diagram: For all years of the survey, it is either in the low demandingness and low effectiveness area (2019, 2021) or in the high demandingness and low effectiveness area (2018, 2020, 2022).
- Conclusion: Visitors seem relatively satisfied with the staff throughout the survey. On the other hand, the weights for this criterion show that the presence of staff has a moderate to low impact on the overall satisfaction of visitors, which seems reasonable since the type of experience of visiting a place like Samaria National Park focuses more on nature and less on any services provided by the park employees.

Local Businesses

- Satisfaction Level: Satisfaction from local businesses was found to be relatively low in 2018 (0.629), 2019 (0.66) and 2020 (0.641) but increased in 2021 (0.741) reaching 0.829 in 2022.
- Weight: The weight ranged between 7.417 and 10.461.
- Demandingness Indices: Visitors were generally not demanding or neutral towards local businesses.
- Effectiveness Indices: Improvements in local businesses would have a moderate impact in absolute terms on increasing overall satisfaction.
- Action diagram: From 2018 to 2021 it is in the Status Quo area indicating low importance and low performance. 2022 moves to the Transfer Resources area, indicating low importance but high performance.
- Improvement diagram: From 2018 to 2020 it is in the high demandingness and high effectiveness area. 2021 moves to the low demandingness and high effectiveness area. 2022 is in the low demandingness and low effectiveness area.
- Conclusion: There was a significant improvement in satisfaction from local businesses over the 5-year period, specifically in 2021 and 2022. At the same time, while in the first 3 years the weight of this criterion was among the lowest compared to the other criteria, in 2021 and 2022 it increased comparatively, showing that local businesses seem to play a greater role in visitor satisfaction, although this role continues to be less important than other criteria.

Resting Areas

- Satisfaction Level: Satisfaction was relatively high, reaching 0.887 in 2022.
- Weight: The weight increased significantly in 2022 (18,575).
- Demandingness Indices: Visitors were not demanding or neutral towards resting areas.
- Effectiveness Indices: Improvements to the resting areas would have a moderate impact in absolute terms on increasing overall satisfaction.

- Action diagram: In 2018 and 2019 the criterion is in the Status Quo area indicating low importance and low performance. In 2020 it is on the border between the Leverage Opportunity and Transfer Resources areas indicating moderate importance and low performance. In 2021 it moves into the Transfer Resources area while in 2022 it is in the Leverage Opportunity area indicating the highest importance and performance.
- Improvement diagram: For the years 2018, 2020, 2021 and 2022 the criterion is in the area of low demandingness and low effectiveness. In 2019 it is in the area of high demandingness and low effectiveness.
- Conclusion: Visitors appear to be consistently satisfied with the location, infrastructure and the condition of the resting areas along the path for all years of the survey. A significant difference is recorded in the weight of this criterion during the 2022 tourist season where it rises to 18,575 and seems to be a very important factor affecting the overall satisfaction of visitors. Again, this may be due to the large number of visitors recorded in this season and the negative consequences that this may have had for visitors along a narrow path, while on the other hand the resting areas along the path offered more space for visitors so that they can enjoy nature.

Price

- Satisfaction Level: Satisfaction from the price showed some fluctuations between the years of the survey.
- Weight: The weight of this criterion is one of the lowest compared to the other criteria.
- Demandingness indices: Visitors were generally not demanding or neutral towards price.
- Efficiency indices: Improvements in price would have a moderate impact in absolute terms on increasing overall satisfaction.
- Action diagram: For all years of the survey it is in the Status Quo area indicating low importance and low performance.
- Improvement diagram: From 2018 to 2021 it is in the area of high demandingness and high effectiveness although in 2019 it was on the border towards the area of low demandingness and high effectiveness. In 2022 it is in the area of high demandingness and low effectiveness.
- Conclusion: Visitor satisfaction with the price of the Samaria National Park entrance ticket appears to be at a good level. At the same time, the low weights calculated for this criterion indicate that visitors do not attach particular importance to the cost in order to enjoy the experience of visiting the Samaria National Park.

Chapter 6
Conclusions: Suggestions for the Future

6.1 Conclusions

Overall, the study shows that Samaria National Park is a very popular destination that attracts a wide range of visitors. The park received consistently positive reviews, and visitor satisfaction remained high despite challenges, such as the COVID-19 pandemic. Based on the results, especially of application of the MUSA method, we see interesting differences between years. We will try to draw some general conclusions here that will help better understand the situation since understanding the criteria that influence visitor satisfaction in protected areas is essential for the sustainable management and conservation of these areas.

An important observation in studies on visitor satisfaction in protected areas is that satisfaction is not a static measure, but an evolving one, shaped by the changing human-nature relationship, visitor profiles and global awareness of environmental issues (Manning, 2011).

For example, in previous decades, physical aspects, such as the availability of clean facilities, accessibility and safety, may have been the main factors determining satisfaction. Recently, however, there has been an increasing emphasis on environmental conservation as a key factor contributing to the overall visitor experience (Coghlan, 2011).

Over time, the criteria that influence satisfaction are subject to dynamic changes caused by a variety of internal and external factors. Internal factors include changes in a place's management practices, the availability and quality of facilities, and the state of natural resources, while external factors include broader societal trends, such as changes in tourist expectations, economic conditions, environmental changes and other exogenous events (McCool, 2009).

Due to the COVID-19 pandemic, there has been a global increase in people's interest in visiting nature, and natural areas have become essential infrastructure

C. Zopounidis, G. Fakotakis, *Multi-Criteria Methodology for the Dynamic Development of Tourist Areas*, Multiple Criteria Decision Making,
https://doi.org/10.1007/978-3-031-95314-9_6

that provides people with outlets and physical and mental benefits (Derks, et al. 2020). The pandemic has increased awareness of health and life risks, leading to a preference for travel options that are safer, more sustainable, and focused on reconnecting with nature as people seek refuge from the stress of urban life, choosing destinations such as national parks, forests, and coastal areas that offer open spaces, clean air, and opportunities for outdoor activities such as hiking, camping, and ecotourism (Zhong, et al., 2021).

The pandemic has significantly changed the criteria that influence tourist satisfaction, with health, safety and hygiene becoming priorities. Tourists have been more concerned about measures such as social distancing and cleaning protocols, while crowding has become a significant negative factor. In addition, the way they perceive safety and risk is now different (Cruz-Milan, 2023).

The survey results from Samaria National Park reveal a significant difference between the years of the study, particularly concerning the concept of safety and its importance to visitors. Before the pandemic, safety in a national park or nature, in general, was primarily associated with protection from natural hazards. Visitors focused on navigating difficult sections of trails to avoid accidents or injuries from slipping, tripping, or falling rocks. They also remained cautious of potential dangers such as snakes and prepared for unpredictable weather conditions or other natural disasters.

However, with the advent of the COVID-19 pandemic, the concept of safety in our daily lives took on a new dimension that was previously unimaginable to such an extent, shifting focus toward health and the prevention of disease transmission. The need for social distancing and avoiding crowds became central to everyday life, accompanied by strict measures such as wearing masks in crowded places, frequent hand disinfection, and maintaining a safe distance from others.

Like all sectors of society, many national parks were forced to adapt to the new conditions, introducing Like all sectors of society, many national parks were forced to adapt to new conditions by implementing necessary hygiene and safety protocols. For instance, facilities such as visitor centers, restrooms, and rest areas were cleaned more thoroughly and frequently. To prevent overcrowding, the number of visitors allowed in popular areas was limited, and certain trails or areas were temporarily closed to ensure compliance with safety measures.

At the same time, technology played a key role in enhancing safety. Many parks began using online platforms to provide information to visitors before their arrival, encouraging them to plan their trips and avoid peak times. Online reservations for park entry and facilities also became more widespread, allowing for better management of visitor flow.

The pandemic has highlighted the importance of mental health, with visits to nature becoming a vital escape from the isolation and stress of lockdowns. Even after restrictions were eased, the lingering effects of the pandemic continued to impact people's well-being.

Overall, the pandemic has reshaped how we perceive safety, including in natural settings, integrating health considerations into the experience (Cruz-Milan, 2023). This shift is reflected in the findings of this research. Based on new data, priorities that once seemed important before the pandemic now matter far less. Visiting areas of natural beauty, away from the hustle and bustle of the city, became synonymous with safety—regardless of the 'natural hazards' that might accompany such places. This trend appears to hold true for visitors to Samaria National Park, where, from 2020 onward, traditional notions of safety became less critical to their satisfaction. In such a setting, safety was simply a given.

Another interesting observation relates to visitors' perception of available time. The amount of time a visitor has when exploring a national park is a significant factor influencing their satisfaction in multiple ways. Sufficient time allows visitors to explore more areas, visit key attractions, navigate various trails, and engage in activities such as observing flora and fauna, photography, or—depending on park regulations—camping, fishing, educational programs, guided tours, and workshops. It also enables them to move at their own pace, take breaks as needed, and enjoy the experience without stress.

The importance of available time for visitors to Samaria National Park is evident in the survey results, with high weights assigned to this criterion across most years, particularly in 2020 and 2021. During this period, visitors generally perceived their available time as sufficient, leading to high evaluations. However, a notable shift occurred in 2022, when the evaluation of available time dropped to its lowest level in the survey's history. This decline coincided with a significant increase in visitor numbers, as 2022 marked the first year many felt the pandemic was behind them. The surge in visitors not only exceeded pandemic-era numbers but also surpassed pre-pandemic levels.

This increase likely negatively impacted visitors' perception of available time. Large crowds led to queues and delays at key facilities such as entrances, restrooms, and information centers, reducing the time visitors could spend on activities. Overcrowding on trails and at major points of interest may have forced visitors to shorten their time at each location to see everything. Additionally, the presence of many others could create stress and discomfort, as visitors might feel pressured to move quickly or seek less crowded areas. The strain on park infrastructure, such as rest areas and restrooms, further exacerbated the issue. Crowded conditions also diminished the sense of calm and relaxation that visitors typically seek in national parks, as well as the opportunity for unique, personal connections with nature—factors crucial to visitor satisfaction.

This conclusion is supported by the fact that the highest visitor satisfaction scores were recorded in 2020 and 2021, when visitor numbers were significantly lower due to the pandemic, and the weight assigned to available time was notably higher. Interestingly, while the 2022 visitor numbers showed a significant increase compared to pre-pandemic levels, the rise may not appear drastic in absolute terms. This raises the question of why the evaluation of available time dropped so sharply in 2022. One possible explanation is that post-pandemic visitors have heightened expectations for relaxation and reconnection with nature, requiring more stress-free time. After the lockdown experience, they may also prioritize comfort and seamless experiences, which are directly influenced by the time available. Additionally, the 2022 visitor numbers may have approached the site's "carrying capacity"—the maximum number of visitors the park can accommodate without negatively impacting the environment or the overall experience. However, this concept has yet to be studied for Samaria National Park.

A third interesting observation concerns the role of resting areas along the route. While resting areas generally influence visitor satisfaction at natural sites, the survey results for Samaria National Park from 2018 to 2021 suggest they played a relatively minor role, as the weights derived from MUSA were moderate to low. However, a significant shift occurred in 2022, when the importance of resting areas increased substantially, ranking first among all criteria. Concurrently, satisfaction with resting areas also reached its highest score compared to previous years.

One possible explanation, when combined with other survey findings, is that the heightened importance of resting areas in 2022 may be linked to visitors' dissatisfaction with the time available to them during their visit. When visitors feel time-pressured, they tend to optimize their experience, and resting areas become crucial for several reasons. They provide convenient spots for quick, efficient breaks, allowing visitors to rest, hydrate, and use restrooms without straying from the trail or significantly disrupting their schedule. This is especially important when fatigue increases due to faster-paced movement.

Strategically placed resting areas also aid in time management. Visitors can plan breaks at designated points, creating a sense of structure and reducing the feeling of being rushed. Additionally, informational signs at resting areas can offer guidance on time requirements between stops, helping visitors navigate Samaria Gorge more efficiently.

From a psychological point of view, resting areas can significantly reduce stress. Positioned at points of high interest, they offer visitors moments of tranquility to admire the view, enhancing their overall experience even under time constraints.

Additionally, resting areas facilitate social interaction. They provide spaces for families and groups to gather, interact, and coordinate activities without wasting time. These areas can also serve as meeting points, allowing groups to regroup and stay together while accommodating individual walking paces. This makes the visit smoother and more enjoyable for everyone.

The significance of this finding is further highlighted by the impact of increased visitor numbers on the overall experience. In overcrowded conditions, such as those observed in Samaria National Park, the role of resting areas becomes even more critical, especially when visitors face time limitations due to park operating rules.

Resting areas allow visitors to temporarily escape crowded trails, offering opportunities to enjoy the natural beauty in peace. They also play a key role in managing visitor flow and reducing trail congestion. Even if one rest area is crowded, visitors can move to the next, distributing crowds more evenly. This improves comfort levels and enhances the overall experience, even during peak visitation periods.

A fourth observation relates to landscape as a criterion influencing visitor satisfaction. Overall, visitors have expressed satisfaction with the landscape during their visits to Samaria National Park. From 2018 to 2021, the MUSA analysis assigned low to moderate weight to this criterion. However, a notable shift occurred in 2022.

While satisfaction with the landscape was slightly lower compared to previous years, it emerged as the third most important criterion based on its weight.

This change may be linked to the increased number of visitors in 2022. Higher foot traffic can lead to environmental degradation, such as litter, soil erosion, and damage to vegetation, which diminishes the quality of the landscape and reduces visitor satisfaction. It can also disturb the park's wildlife, causing animals to avoid trails and reducing visitors' chances of encountering them. Additionally, overcrowding at popular spots can disrupt the sense of peace and relaxation that many visitors seek, preventing them from fully enjoying the scenery.

Finally, when a natural area becomes overcrowded, it may fail to meet visitors' expectations. Promotional materials often depict pristine, untouched landscapes, creating high expectations. If the reality falls short due to overcrowding or environmental strain, it can negatively impact the overall visitor experience.

6.2 Suggestions for the Future

6.2.1 Carrying Capacity Study

The concept of carrying capacity is essential for the management and conservation of national parks and natural areas. It refers to the maximum number of visitors a site can accommodate without causing significant environmental degradation, overloading infrastructure, or negatively impacting the visitor experience. Determining and applying carrying capacity is critical to balancing environmental protection with visitor enjoyment. However, such a study has not yet been conducted for Samaria National Park. The results of this research suggest that the high number of visitors, particularly in 2022, may have affected satisfaction levels and potentially approached the park's carrying capacity.

Carrying capacity can be divided into four categories, each addressing different aspects of park management and protection:

1. Ecological Carrying Capacity: This refers to the level of visitor use an ecosystem can sustain without significant degradation. It focuses on minimizing impacts on flora, fauna, soil, water resources, and other natural features (Manning, 2007).
2. Physical Carrying Capacity: This relates to the maximum number of visitors that park infrastructure—such as trails, parking lots, rest areas, and restrooms—can support (Eagles et al., 2002).
3. Social Carrying Capacity: This refers to the number of visitors beyond which satisfaction begins to decline due to overcrowding (Wagar, 1964).
4. Managerial Carrying Capacity: This concerns the ability of park management to handle visitor numbers based on available staff, financial resources, and other operational capabilities (Manning, 2007).

Several factors influence carrying capacity. Parks with fragile ecosystems, rare species, or unique landscapes typically have lower visitor capacities (Manning, 2007). The types of activities visitors engage in—ranging from low-impact to highly intrusive—also play a role (Eagles et al., 2002). Additionally, the availability and quality of infrastructure determine how many visitors can be accommodated safely at any given time (Wagar, 1964). Finally, management practices, such as visitor monitoring and infrastructure maintenance, are critical in maintaining a sustainable balance (Stankey et al., 1985).

Calculating the carrying capacity for a national park is a complex task. Natural ecosystems are dynamic, and the impact of visitor numbers can vary significantly from year to year (Manning, 2007). Balancing environmental protection with visitor numbers and satisfaction is crucial (Eagles et al., 2002). However, this effort often faces pressure from stakeholders who prioritize increasing visitor numbers to boost the local economy (Farrell & Marion, 2002) and may resist restrictions. Collecting and evaluating data on visitor numbers, behavior, and their environmental impacts is challenging but essential for accurately determining carrying capacity (Riebau, 1994).

Implementing a management plan based on carrying capacity also presents several challenges. Continuous monitoring of visitor numbers, behavior, and environmental impacts is necessary. It may require defining specific use and protection zones to reduce pressure on vulnerable areas (Eagles et al., 2002) and ensuring ongoing infrastructure improvement and maintenance (Wagar, 1964). Additionally, any restrictions on visitor use must be accompanied by clear communication to visitors, encouraging behavior that minimizes negative impacts and supports conservation efforts (Farrell & Marion, 2002).

Given these complexities, the concept of carrying capacity for a site like Samaria National Park requires evaluating and balancing competing factors. Multi-criteria analysis is an ideal tool for this purpose, as it enables a comprehensive assessment that incorporates environmental, social, economic, and infrastructure factors. This approach maintains the delicate balance between protection and visitor use.

Multi-criteria analysis allows for the inclusion and weighting of criteria based on stakeholder preferences and priorities. It can handle both quantitative data (e.g., visitor numbers, pollution levels) and qualitative data (e.g., visitor satisfaction, aesthetic value), providing a structured framework to assess the relative importance of each criterion. It also supports the modeling of different management scenarios and their potential impacts on carrying capacity. This predictive capability empowers park managers to make informed decisions and develop strategies for long-term sustainability.

Furthermore, multi-criteria analysis promotes stakeholder participation, ensuring transparency and clarity in the decision-making process. By involving stakeholders in defining criteria and weighting their importance, the analysis can reflect the preferences of the local community. The ultimate goal is to generate clear, actionable insights that support effective decision-making and strategies for maintaining or enhancing carrying capacity.

6.2.2 *Use of Sub-Criteria and Additional Criteria*

A limitation of the current research on Samaria National Park is its focus on main criteria for analyzing visitor satisfaction, without incorporating sub-criteria. Including sub-criteria is highly beneficial, as it enables a more detailed and precise analysis of the factors influencing the visitor experience. This approach provides valuable insights for improving park management and enhancing overall visitor satisfaction.

The use of sub-criteria allows researchers to break down each main criterion into specific components. This offers a clearer understanding of the individual aspects that contribute to visitor satisfaction or dissatisfaction. As a result, stakeholders can identify and address specific areas needing improvement, leading to more targeted and effective interventions.

Based on the criteria used in this study, potential sub-criteria could include:

Sub-criteria for Tour Guides

- Knowledge and Expertise: The depth of the guides' knowledge about the park, including its history, flora, and fauna.
- Communication Skills: The ability of guides to convey information clearly and engage visitors effectively.
- Friendliness and Approachability: The guides' attitude, willingness to answer questions, and interact with visitors.
- Adaptability of Tours: Flexibility in tailoring tours to the interests and needs of different visitor groups.
- Emphasis on Safety: The guides' ability to manage emergencies and ensure visitor safety during tours.

Sub-criteria for Signage

- Clarity and Legibility: Factors such as font size, color contrast, and language used on signs.
- Information Content: The quality and relevance of information provided, including details on history, flora, fauna, and safety.
- Placement: The strategic positioning of signs to ensure maximum visibility and usefulness for visitors.
- Multilingual Availability: The inclusion of signs in multiple languages to accommodate visitors of different nationalities.
- Maintenance: The condition, upkeep, and cleanliness of signs to ensure they remain legible and visually appealing.

Safety Sub-Criteria

- Emergency Services: Availability and response time of doctors and rescuers in emergencies.
- Safety Infrastructure: Presence of barriers, railings, fire stations, assembly points, escape routes, and guard posts to prevent accidents.
- Weather Information: Availability of real-time weather updates and warnings.

- Visitor Information: Provision of information on potential hazards and safe practices within the park.
- Incident Reporting: Ease of reporting accidents or suspicious activities and the efficiency of resolving them.

Available time Sub-Criteria

- Visit Options of Different Durations: Availability of trails and activities tailored to various schedules.
- Entry and Exit Efficiency: Waiting times for entry and exit, including ticketing and security checks.
- Activity Scheduling: Information on peak times and suggested itineraries to help visitors plan their visit effectively.
- Time Management Information: Availability of estimated time requirements for various activities and trails.

Landscape Sub-Criteria

- Points of Interest: Accessibility and quality of viewpoints and trails that showcase the park's scenery.
- Alternative Trails: Variety of trails offering diverse landscapes.
- Environmental Quality: Cleanliness, preservation of natural ecosystems, and minimization of human impact.

Staff Sub-Criteria

- Friendliness and Politeness: Staff demeanor and willingness to assist visitors.
- Professionalism: Adherence to rules and professional conduct.
- Staff Availability: Number of staff and ease of locating them throughout the park.
- Problem Solving: Efficiency in addressing visitor complaints and issues.

Local Business Sub-Criteria

- Quality of Products and Services: Standard of services and products offered by local businesses (e.g., restaurants, souvenir shops, hotels).
- Variety: Diversity of products and services available to visitors.
- Sustainable Practices: Adoption of environmentally friendly and sustainable practices by local businesses.

Resting Area Sub-Criteria

- Availability: Number and distribution of rest areas throughout the park.
- Infrastructure: Quality of seating, shading, weather protection, drinking water, and restroom facilities.
- Condition: Maintenance and cleanliness of rest areas.
- Location: Strategic placement of rest areas at points of interest.
- Accessibility: Ease of access for all visitors, including those with disabilities.

Price Sub-Criteria

- *Admission Price*: Visitors' perception of the admission fee in relation to the overall experience offered.
- *Additional Costs*: Extra expenses visitors may incur, such as parking fees, guided tours, activities, or transportation to and from the park.

- *Discounts and Offers*: Availability of discounts, packages, and special promotions.
- *Comparative Pricing*: Comparison of prices with similar attractions or parks.

By incorporating these sub-criteria into a multi-criteria analysis, future research can gain a more detailed understanding of visitor satisfaction at Samaria National Park. This approach not only helps identify specific areas for improvement but also enhances overall park management and planning to better align with visitor expectations. Additionally, further criteria can be explored and integrated as needed.

6.2.3 Using Sentiment Analysis

Sentiment analysis, also known as opinion mining, is a process that uses Natural Language Processing (NLP), text analysis, and computational linguistics to identify and extract subjective information from text data. Its primary goal is to determine the sentiment expressed in a piece of text, categorizing it as positive, negative, or neutral (Liu, 2015).

Various techniques are used to perform sentiment analysis:

- Rule-based approaches: These rely on predefined rules, including lists of words and phrases associated with positive or negative emotions, to determine sentiment.
- Machine learning approaches: These involve training a model using a dataset where each text is pre-categorized by sentiment (Pang et al., 2002).
- Hybrid approaches: These combine rule-based and machine learning methods for greater accuracy (Liu, 2015).

Sentiment analysis has applications across multiple fields. Companies use it to assess customer satisfaction and improve products or services based on feedback. In market research, it helps gauge public opinion about products, brands, or services, aiding in the development of marketing strategies. Organizations also monitor social media to understand public sentiment on various issues or events. In politics, it can analyze citizens' opinions about candidates and policies (Pang & Lee, 2008).

For Samaria National Park, sentiment analysis can be a powerful tool for collecting visitor satisfaction data. By analyzing visitor comments from sources such as online reviews, social media posts, and survey responses, park management can gain deeper insights into visitors' experiences and emotions, complementing data from questionnaires and interviews. For example:

- Online reviews: Platforms like TripAdvisor and Google Reviews can be analyzed to identify common themes in visitor feedback.
- Social media: Posts on platforms like Twitter, Instagram, and Facebook can be examined using tools like VADER (Hutto & Gilbert, 2015) to understand visitors' sentiments.

For instance, frequent complaints about congestion, infrastructure issues, or cleanliness problems can highlight areas needing improvement. Conversely, positive feedback about staff helpfulness or scenic beauty can identify strengths.

Sentiment analysis can also be integrated with multi-criteria analysis. By identifying which park features receive positive or negative feedback, sentiment analysis can help prioritize criteria for improvement. For example, if visitors express strong positive sentiments about the landscape but negative sentiments about restroom facilities, multi-criteria analysis can evaluate the importance of these factors and guide decision-making to enhance visitor satisfaction.

6.2.4 Continued Satisfaction Research

Continued research into visitor satisfaction at Samaria National Park is essential for sustainable park management, environmental protection, and enhancing the visitor experience. Below are key reasons why ongoing research is vital:

1. Evolving Visitor Preferences and Expectations:
   Visitor preferences are not static; they change over time due to demographic shifts, technological advancements, societal trends, and unforeseen events like the pandemic. Ongoing research helps stakeholders and park management monitor these changes and adapt accordingly. For example, modern visitors may seek more interactive and educational experiences, digital engagement through apps or virtual tours, or access to lesser-explored areas of the park. Understanding these evolving needs ensures the park continues to meet or exceed visitor expectations.
2. Identifying Areas for Improvement:
   Visitor feedback provides invaluable insights into specific strengths and weaknesses. Continuous research enables targeted interventions to address issues and enhance overall satisfaction. For instance, if visitors frequently highlight the need for more rest areas or better waste management, this feedback can guide infrastructure upgrades and operational improvements.
3. Promoting Sustainable Tourism:
   Sustainable tourism is a cornerstone of modern national park management, aiming to minimize negative impacts while maximizing benefits. Ongoing satisfaction research helps develop strategies that promote sustainable behaviors. For example, understanding visitor perceptions of crowding can inform measures to distribute visitors more evenly, reducing pressure on ecosystems and improving the experience for all.
4. Balancing Conservation and Recreation:
   Samaria National Park is renowned for its biodiversity and pristine landscapes. Continuous research helps balance conservation goals with recreational use. By assessing visitor awareness and attitudes toward conservation, stakeholders can design educational programs that foster appreciation and respect for the environment. This not only enhances the visitor experience but also supports preservation efforts.

5. Data-Driven Decision-Making:
   Effective park management relies on data-driven decisions. Ongoing research provides a steady stream of data to inform operational and strategic planning. For example, understanding seasonal variations in visitor satisfaction can improve management during peak months. It also allows for the evaluation of new initiatives, such as facilities or programs, and adjustments based on visitor feedback.
6. Economic Benefits:
   Visitor satisfaction is closely tied to the park's economic success and its impact on the local economy. Satisfied visitors are more likely to return, recommend the park, and spend more during their visit. Continuous research ensures the park delivers a high-quality experience, boosting its reputation and economic contribution. Additionally, understanding visitor preferences can guide promotional efforts to attract new visitor segments, further expanding the park's economic impact.
7. Enhancing Cultural and Educational Value:
   Samaria National Park holds significant cultural and historical value. Visitor satisfaction research can reveal opportunities to enrich its educational and cultural offerings. For example, surveys may show a desire for more information about the park's heritage or guided tours exploring local history and traditions. Addressing these desires can deepen visitors' connection to the park.
8. Addressing Challenges:
   The park faces challenges such as climate change, natural disasters, and fluctuating visitor numbers due to events like the pandemic. Ongoing research helps stakeholders develop strategies to maintain high satisfaction levels even during difficult times. By understanding how various factors affect the visitor experience, management can create contingency plans to address these challenges effectively.

Ongoing visitor satisfaction research is crucial for maintaining and enhancing the quality of the visitor experience, supporting sustainable tourism, and ensuring the long-term conservation of Samaria National Park's natural and cultural resources. By continuously collecting and analyzing visitor feedback, park management can adapt to changing expectations, improve facilities and services, promote environmental protection, and deliver economic benefits to the local community.

This approach is not limited to Samaria National Park; its methodologies and findings can be applied to other protected areas worldwide. By doing so, natural areas can remain cherished destinations for future generations, balancing conservation goals with economic viability and visitor satisfaction.

References

Coghlan, A. (2011). Linking natural resource management to tourist satisfaction: A study of Australia's Great Barrier Reef. *Journal of Sustainable Tourism, 20*(1), 41–58.

Cruz-Milán, O. (2023). *Loyalty in the time of COVID-19: A review of the literature in tourism destination settings* (Vol. 14, p. 1119737). Frontiers in Psychology.

Derks, J., Giessen, L., & Winkel, G. (2020). COVID-19-induced visitor boom reveals the importance of forests as critical infrastructure. *Forest Policy and Economics, 118*, 102253.

Eagles, P. F. J., McCool, S. F., & Haynes, C. D. (2002). *Sustainable tourism in protected areas: Guidelines for planning and management*. IUCN Gland.

Farrell, T. A., & Marion, J. L. (2002). The protected area visitor impact management (PAVIM) framework: A simplified process for making management decisions. *Environmental Management*.

Hutto, C. J., & Gilbert, E. (2015). VADER: A parsimonious rule-based model for sentiment analysis of social media text. In *Proceedings of the 8th international conference on weblogs and social media, ICWSM 2014*.

Liu, B. (2015). *Sentiment analysis: Mining opinions, sentiments, and emotions*. Cambridge University Press.

Manning, R. E. (2007). *Parks and carrying capacity: Commons without tragedy*. Island Press.

Manning, R. E. (2011). *Studies in outdoor recreation: Search and research for satisfaction*. Oregon State University Press.

McCool, S. F. (2009). Constructing partnerships for protected area tourism planning in an era of change and messiness. *Journal of Sustainable Tourism, 17*(2), 133–148.

Pang, B., & Lee, L. (2008). Opinion mining and sentiment analysis. *Foundations and Trends in Information Retrieval, 2*(1–2), 1–135.

Pang, B., Lee, L., & Vaithyanathan, S. (2002). Thumbs up? Sentiment classification using machine learning techniques. In *Proceedings of the ACL-02 conference on empirical methods in natural language processing* (Vol. 10, pp. 79–86).

Riebau, A. R. (1994). Wilderness monitoring and data management. *Environmental Software, 9*(4), 227–232.

Stankey, G. H., Cole, D. N., Lucas, R. C., Peterson, M. E., Frissell, S. S., & Washburne, R. F. (1985). *The limits of acceptable change (LAC) system for wilderness planning*. USDA Forest Service General Technical Report.

Wagar, J. A. (1964). The carrying capacity of wildlands for recreation. *Forest Science Monographs, 7*, 1–23.

Zhong, L., Sun, S., Law, R., & Li, X. (2021). Tourism crisis management: Evidence from COVID-19. *Current Issues in Tourism, 24*(10), 1233–1246.

The manufacturer's authorised representative in the EU is Springer Nature Customer Service Centre GmbH, Europaplatz 3, 69115 Heidelberg, Germany. If you have any concerns regarding our products, please contact ProductSafety@springernature.com

Printed and bound by CPI Group (UK) Ltd, Croydon, CR0 4YY
30/09/2026
02225577-0005